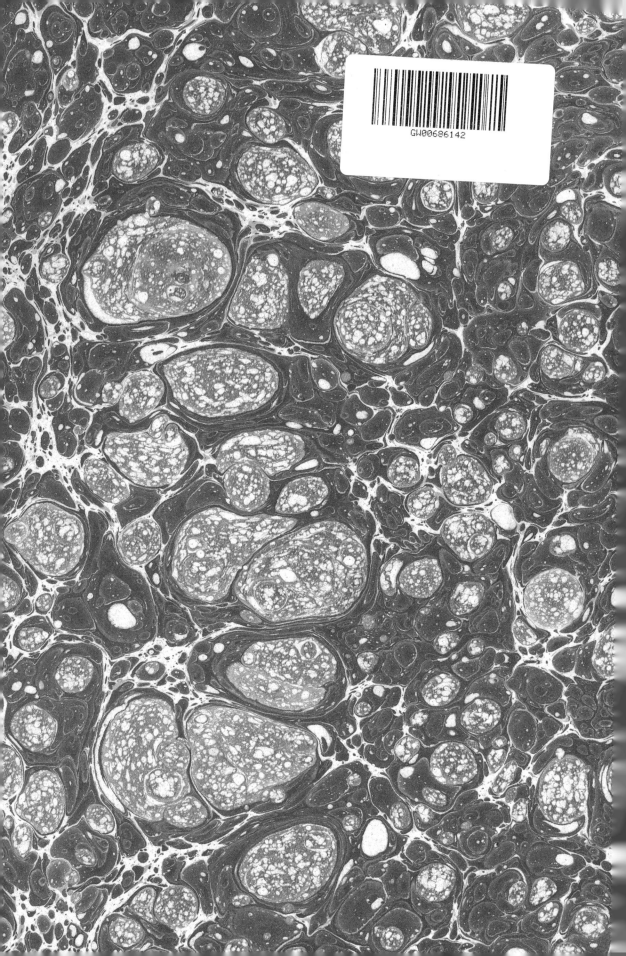

Dublin 22/11/02

MANFRED GRIEHL/JOACHIM DRESSEL

LUFTWAFFE COMBAT AIRCRAFT

Manfred Griehl/Joachim Dressel

LUFTWAFFE COMBAT AIRCRAFT

Development • Production • Operations
1935-1945

Translated from the German by Don Cox

Schiffer Military/Aviation History
Atglen, PA

Dust jacket artwork by Steve Ferguson, Colorado Springs, CO

EGYPT'S HELL FIRE
The dust jacket depicts a flight of RAF Hurricanes intercepting Messerschmitt Bf 110D Zerstörers of 8./ZG 26 (Zerstörergeschwader) on a strafing mission near Halfaya, Egypt in late April, 1941. After two years of fighting for the strategic coastal road through the hills near Halfaya, veterans aptly named it "Hell Fire Pass." Famous ZG 26 had reprieved its rather dismal combat record in the Battle of Britain in 1940 by racking up an impressive number of kills in the subsequent winter campaigns over southern Russia, and particularly in Greece which had not fallen to their Italian allies. As the Italians further struggled in the conquest of Egypt, III/ZG 26 advanced from the Aegean area in the third week of April, 1941, to reinforce the critical venture in North Africa. In time, the RAF brought in more of ther cursed Hurricanes and Spitfires plus the rugged P-40s supplied by the U.S. By year's end, the Zerstörers were once again forced out of the air supremacy battle. While flying convoy escort duty and ground support missions, III/ZG 26 suffered devastating losses in the year long battle for Tunisia. ZG 26 retired to Germany in early 1943 to join the Reich defense forces, only to be wiped out the following year by the invading Allied air force.

Translated from the German by Don Cox.

This book was originally published under the title,
Die Deutschen Kampfflugzeug im Einsatz 1935-1945,
by Podzun-Pallas Verlag.

Copyright © 1994 by Schiffer Publishing Ltd.
Library of Congress Catalog Number: 94-66971.

Printed in the United States of America.
ISBN: 0-88740-683-1

We are interested in hearing from authors with book ideas on related topics.

Published by Schiffer Publishing Ltd.
77 Lower Valley Road
Atglen, PA 19310
Please write for a free catalog.
This book may be purchased from the publisher.
Please include $2.95 postage.
Try your bookstore first.

Contents

Introduction   7

Pilot Training   8
Training for an Operational Pilot   8
Training Aircraft of the Luftwaffe   9
Training and Auxiliary Units   12

Testing   27
From Proposal Request to Prototype   27
Factory Testing and the K.d.E.   29
The Luftwaffe's Erprobungsstaffeln   39

Production   51
The German Aircraft Manufacturers and their Most
Important Products   51
Dispersal of Production   55
Aircraft Equipment and Armament   56
Drop Ordnance and Aerial Torpedoes   61
Barrel and Rocket Armament   70

The Luftwaffe in Action   115
Units and Organization of the Luftwaffe   115
From Blitzkrieg to the Defensive   117
New Missions and Tactics 1944/1945   157

Photo Credits   194

Introduction

Up to now a wealth of books and manuscripts has been published on the German Luftwaffe. This current book is a photo essay which, with its portrayal of diverse scenes, furnishes the model builder with valuable stimulus for building dioramas and at the same time provides interesting photographs and data. This book also gives answers to a number of questions regarding construction, armament and operations of the variously equipped Luftwaffe units.

We are particularly grateful to the following individuals and organizations who have made this book possible: Messrs G. Aders, N. Arena, U. Balke, C.E. Charles, E. Creek, H.P. Dabrowski, Dornier GmbH, K. Francella, Dr. Wustrack of Flughafen Frankfurt AG, B.A. Filley, H. Grimminger, P. Heck, Dr. A. Hiller, Henschel-Werke, P.K. Herrendorf, Dipl.-Ing. K. Kössler, B. Lange, R.P. Lutz Jr., J. Meier of the Deutsches Aero Club e.V., Messerschmitt-Bölkow-Blohm, MTU, van Mol, F. Müller-Romminger, P. Petrick, W. Pervesler, J. Prowan, R. Olejnik (deceased), G. Ott, W. Radinger, H. Riediger, Ch. Regel, H. Schliephake, Ph. Schreiber, F. Selinger and R. Smith, as well as Messrs D. Herwig of the Deutsches Studienbüro für Luftfahrt and B. Lange, H. Sander, H.H. Stapfer, H. Spork, R. Schirer, G. Schlaug, H. Thiele, O. Thiele, J. Menke, H.J. Nowarra, F. Marshall, H.J. Meier and F. Trenkle. Also Messrs Dipl.-Ing. Pöltisch (deceased), B. Wittigayer, G. Wegmann, the Wehrbereichsbibliothek IV and the Zentralbibliothek der Bundeswehr, as well as Herr Dipl.-Ing. Zucker (deceased).

Manfred Griehl
Joachim Dressel

Pilot Training

Training for an Operational Pilot

Beginning in 1921, the Weimar Republic saw the formation of numerous independent flying clubs which soon merged to become the Deutscher Luftfahrt Verband e.V. (DVL). Initially, this organization issued testing requirements for the A-, B- and C- licenses for glider pilots. Later, however, powered flight training began, albeit only in secret. An example of this was the "Deutsche Verkehrsfliegerschule GmbH" in Schleissheim, which played an important role in follow-on training for those completing their B1 qualifications. At the same time advanced training – up to C2 classification – was also being carried out in Braunschweig. Maritime aircraft training was conducted in Warnemünde and List. In 1931 a secret Fliegerwaffenschule, or Pilot's Weapons School, was established in Braunschweig, since Lipetsk in the USSR did not have nearly the capacity to provide adequate training.

The varying "schools" flew a true menagerie of aircraft, from the Arado S Ia, Fieseler and Raabkatzenstein biplanes, to the Udet U 12, Albatros L 102 biplane, Junkers F 13 and A 48, Dornier Merkur, and the Klemm L 20, 25 and 26. Then there were the assorted Heinkel types and finally, the Fw 44 "Stieglitz."

After 30 training hours the student pilot received the intermediate license, once he had completed his proficiency aerobatic loop as well as three landings without error. In order to be awarded "basic certification" (A1) it was necessary to complete an altitude flight to 2000 m and a 300 km triangular course. A2 certification, on the other hand, was for aircraft of at least two seats. Usually this was followed by the award of the B1 or B2 certification. The prerequisite, however, was that the pilot could show that he already had 3000 km flight experience and had completed a 600 km triangular course in nine hours. In addition 50 training flights in B1 aircraft as well as an altitude flight to 4500 m were required. On the practical portion of the test, the young pilot must conduct three precision landings, two more night landings and a night flight of at least 30 minutes. B2 certification was even more difficult; minimum prerequisites were flight experience of 6000 km, 3000 of which were to have been completed on B1-class aircraft. 50 additional training flights were required in B2 aircraft in addition to several difficult night landings.

Those pilots designated to fly the large, multi-engine aircraft were eventually given the C1 certification. This certification required that the pilot had already flown 20000 km in B-class aircraft, 6000 of which he was the responsible pilot on board. In addition, the training program called for 30 flights in C1 aircraft as well as displaying general knowledge in the area of aviation communication. The C2 training, in turn, built upon the foundation of the

C1 certification. Aside from another 30 training flights the advanced pilot demonstrate his abilities over several 800 km triangular flight routes, a night flight of 200 km length and two flights with one engine shut off.

However, even upon completion of the above-named flight training it was only possible to receive the so-called "basic certification." For special certification 50000 km was the basic prerequisite. The KI and KII aerobatic licenses could be issued after 50 and 100 flight hours, respectively, and were based upon demonstration of abilities in the A to C aircraft classes.

As early as the autumn of 1935 Germany once again had well equipped training units, when on 1 April the Kommando der Schulen assumed duties. In addition to the pure A2 schools and the seven B2 training facilities, four C2 schools were also set up. In Brandis and Celle instrument flying schools were established and in Schleissheim near Munich an aerobatic and fighter training center was opened. Furthermore, each Luftkreis (air district) saw the organization of a pilot's replacement division and, in 1936, five air warfare schools were founded. The Reischschulen für den Motorflugsport (Reich Schools for Powered Sport Flying) served to enhance the military training capacity.

Training Aircraft of the Luftwaffe

The division of the training aircraft initially corresponded to the pilot certification program and could easily be distinguished by reading the civilian "D" code designation:

Class	Persons	Weight(kg)	Engines	Designation
A1	1-2	to 500	1	D-YAAA to YZZZ
A2	1-3	to 1000	1	D-EAAA to EZZZ
B1	1-3	to 2500	1	D-IAAA to IZZZ
B2	4-6	to 2500	1	D-OAAA to OZZZ
C1	6	over 2500	1	D-UAAA to UZZZ
C2	6	over 2500	2+	D-AAAA to AZZZ

Only a few sub-groups deviated from this. For example, a C2-class aircraft could be operated with fewer than six men on board. However, in all cases the requirements called for a radio-telegraphy set (FT) and multiple engines parallel to the aircraft's lateral axis. During the war this classification was changed on several occasions, and in 1944 – due to the systematic increase in weight – was as follows:

Weight in kg for

Class	Landplanes	Seaplanes	Persons	Aircraft Types
A1	to 500	to 600	1	Erla 5, Bü 180
A2	500-1000	600-2200	1-3	Bü133, Fw 44, He 72
B1	1000-2000	2200-5500	1-4	Ar 66, Fh 104, Go 145
B2	2500-5000	2200-5500	1-8	Fw 58, Junkers W 34
C	over 5000	over 5500	varied	Ju 52, He 111, Fw 200

A "pupil" of the Röhn-Rossiten-Gesellschaft seen just after lifting off. Nearly every hopeful pilot began his basic training in these simple gliders.

The Grunau firm's "Baby II B" was probably the most well-known of the gliders. It was often exported and flew long after the war. The design had a glide ratio of 1:17. With a wingspan of 13.57 m and a length of 6.09 m the sailplane possessed excellent maneuverability.

It should be noted that this classification was used for nearly all operational aircraft in the German Luftwaffe.

The inventory situation with regards to training aircraft was extremely tenuous as the war drew to a close, since on the one hand serious losses had occurred due to breakdowns and crashes and on the other hand the industry was heavily involved with producing combat aircraft – primarily the Fw 190 and Bf 109. At a conference on 27 Oct 1944 it was decided that the variety of training aircraft would be reduced dramatically. The production of the Bü 181 would be increased to a monthly output of 180 airplanes as quickly as possible. The Ar 96 was to be replaced by the Ar 396 (combination design). The Si 204 was also to make use of wooden parts in its construction. Furthermore, it was ordered that the Bf 109 and the Fw 190 would be produced in trainer versions in the future and that a special training aircraft would be developed for the Staffeln equipping with the He 162 at the beginning of 1945. Shortly before the end of the war a handful of Me 262 A-1a fighters were modified to two seat trainers, designated Me 262 B-1a. Of this batch, however, around five were converted into night fighters in the spring of 1945.

Training and Auxiliary units

At the end of 1938 the inventory of the A/B schools generally looked so:

A2/Land: Albatros L101, Bü 131/133/181, Fw 44, He 72 and Kl 25 to 35
B1/Land: Ar 65/66/68/96, Fw 56, Go 145, He 45/46/51 and Bf 108
B1/Sea: He 42W/60W and He 114
B2/Land: Fw 58, He 70, Junkers F 13/W 33/W 34 and Si 204
B2/Sea: Junkers W 33W and W 34 W

In the summer of 1939 the OKL, in agreement with the RLM, determined that each A2 school was to operated 45 trainers, each B1 school 21 aircraft and each B2 school 30 aircraft – at least on paper.

The C-class pilot training schools assumed continued flight training for those pilots with valid B2 certifications. They provided training on heavy, multi-engine models suitable for demonstrating a deeper knowledge of navigation, night and poor weather flying. For instrument flight training, a pilot was required to have a "supplemental Luftwaffe pilot's license" which included the C1 certification. The C schools initially had a similar amount of variety as that found in the A2 to B2 schools:

C2/Land: Do 11/13/23, Do 17, Ju 52/86, later He 111 and Ju 88
C2/Sea: Do 15 "Wal", Do 18, He 59 and He 115

Altogether there were a total of 22 C schools which – as long as it was in operation – fell under the jurisdiction of the "Höheres Kommando der Blindflugschule" (Senior Command of Instrument Flight School) in Berlin. During the course of the war nearly every school was given training and obsolete combat aircraft captured during the various Blitzkriegs. Instrument training schools 1 to 11 not only trained pilots of the Luftwaffe's Kampfgeschwader (bomber units), but also were involved in providing instruction for pilots of the single and multi-engine night fighter units. For a short time there were also JGr. 112 and "Blindflugschule Schleswig." For instrument flight training many aircraft were fitted with canopies painted over at the unit level, such as often could be seen on the Go 145. Generally, however, the most common types flown were the Ju 52 and Ju 88. Blindflugschule 6 in Wesendorf even operated the modern Ju 290 heavy bomber for a few months in order to provide long range bomber crews with near-real combat experience. The ten A/B schools listed initially carried the burden of training in this area:

Flight School	Location of Training	Disbanded/Redesignated
FFS A/B 1	Görlitz and Bautzen	October 1944
FFS A/B 2	Hannover, Deblin and Warsaw	August 1944
FFS A/B 3	Cottbus, Brünn and Güben	September 1944
FFS A/B 4	Nuremberg, Prague, Göppingen and Odergrund	November 1944
FFS A/B 5	Seerappen, Quackenbrück and Augsburg	February 1945
FFS A/B 6	Danzig and area	Summer 1942
FFS A/B 7	Plauen and Chemnitz	January 1945
FFS A/B 8	Marienbad and Grafenwöhr	October 1940
FFS A/B 9	Grottkau, Pardubitz and Weimar	February 1945
FFS A/B 10	Waremünde, Rostock and Güstrow	February 1945

Moreover, even before the war there were the pilot training schools of the various Fliegerausbildungsregimenter(FAR), e.g. Schule 11 based first in Schönwalde, then Gardelegen and finally Neukuhren. A series of additional A/B schools sprang up at the end of 1939, such as A/B 113 and thirteen more, some of which were stationed in the occupied areas.

The Bücker Bü 131 provided the A-2 class training for the young pilots. In the background is a Klemm Kl 35.

Pilot training with the Kl 35 in Fürth. Entry into B class training followed the presentation of the A class license.

The most well-known representative of this group was the B-1 class trainer known as the Focke-Wulf Fw 56 "Stösser." At the beginning of the war each school possessed the following aircraft: 45 class A-2 (Bü 131, Bü 133, Bü 181, Fw 44, He 72, Kl 32, Kl 35, etc.), 21 class B-1 (Ar 65, Ar 76, Ar 96, Fw 56, Go 145, He 42, He 60, He 51, Bf 108, etc.) and 30 B-2 class (Fw 58, He 70, Ju W 34 sea- and landplanes, Si 204, etc.).

Formation flying was also on the training schedule during the approximately 12 months of instruction. Here it's being conducted using the Arado Ar 96.

This Arado Ar 68 was used for B-1 class training in southern Germany.

Several He 60 C seaplanes were put to use in the role of B-1 class maritime flight training at the Flugzeugführerschule See in Stettin.

Captured enemy aircraft could often be found operating in the training role alongside their German counterparts. This is a North American NA-64, which was produced in France under license beginning in 1939. After France's capitulation this type was transferred to the Dienststelle Chef Ausbildungswesen (Office of the Chief of Training) and finally to the AB school for fighter pilot preparatory training.

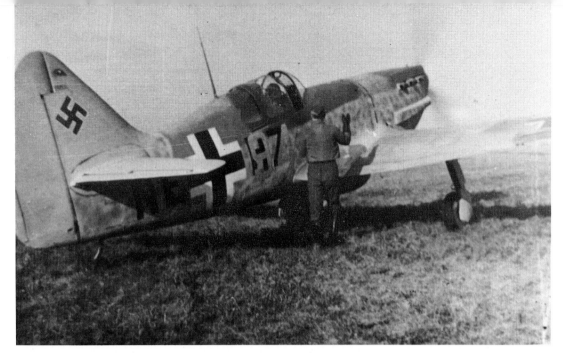

Numerous Dewoitine D 520s were operated in 1944 by JG 101 in Pau as fighter pilot basic trainers. With an 850 hp Hispano-Suiza 12 cylinder inline engine this aircraft reached a maximum speed of 535 kmh. At the outbreak of the war, this aircraft type was considered to be one of the most modern fighters.

The previously standard fighter of Czechoslovakia, the Avia BH 534/IV, was utilized as a K type trainer at various pilot training schools. This accident occurred near the well-built airfield at Prague-Ruszin.

The MS 230, in production up until the early summer of 1944, was used by many sailplane Gruppen from the end of 1940 onward. The aircraft was manufactured by Morane-Saulnier in Puteauz near Paris and served admirably as a towplane for gliders.

The Bf 109 G-12 was converted into a dual-control trainer from the Bf-109 G-1, G-5, B-6 and G-14. The model for these conversions was Werknummer 14130 (DF+CJ). Initial operations commenced in May of 1944 with JG 101 to JG 112 as well as with Nahaufklärungsgeschwader 102.

The Bf 109 G-4 was the next step following conversion training from the G-12/R-3s parked here. The radio equipment usually consisted of an FuG 16 Z with an intercom device.

A Focke-Wulf 190 S-8 with a BMW 801 D-2 engine at the Altenburg airfield in September of 1944. Due to weight constraints armament consisted of only two fuselage mounted MG 17 machine guns.

The Junkers W 34, fitted with a BMW 132 A engine, was utilized in many roles, e.g. for communications training as well as blind flying and navigation training.

The Focke-Wulf Fw 58 C, as a B-2 type, also served as a trainer for comms and instrumentation flying. Note the unusual colors of the fuselage code PW+AM.

Powered by two Renault 6020/21 engines of 240 hp each and with counter-rotating propellers, this Caudron 445 spent its service with Luftflottenflugbereitschaft 3.

Flying accident involving a Junkers Ju 52, DD+GH, belonging to Flugzeugführerschule FFS(C).

Foreign aircraft also found use as C-2 type trainers, as illustrated by this MB 200 with Walter K-14 engines.

After the introduction of the P series, the He 111 J was pulled out of the bomber units and also utilized as a capable C-2 trainer aircraft.

The former Junkers Ju 88 A-6, recognizable by the cable deflectors on the nose, was converted into the Ju 88 A-16 auxiliary trainer with dual controls.

Testing

From Proposal Request to Prototype

As a rule the aircraft manufacturers only produced airplanes upon issuance of a specific contract by the Technisches Amt (GL, Technical Office) – and with the approval of Department C-E of the Reichsluftfahrtministeriums (RLM, the Reich Ministry of Aviation). Guidelines for usage were established first, along with a tentative plan for the number to be produced. Department C II of the Technisches Amt then began to arrange the request for bids on the project, which was then sent to the various manufacturers by Department C IV (Verwaltung, or Administration). The proposals submitted by these companies were examined in detail by the Technisches Amt not only with regards to their developmental maturity, but also in view of their resulting costs, production potential and material requirements. In addition to Amstgruppe Entwicklung von fliegenden Gerät (GL/C7, Office for Aircraft Development) and Amtsgruppe Beschaffung (GL/C-B, Office for Procurement) – along with the Kommandeur der Erprobungsstellen, there were 22 other technical departments which came under the direction of the Chief of the Technisches Amt. Within the framework of overall aircraft and weapons development, responsibility was broken down as follows:

Branch	
Aircraft	GL/C-E2
Engines	GL/C-E3
Communications and Radio Navigation	GL/C-E4
Equipment	GL/C-E5
Guns/Cannons	GL/C-E6
Ordnance	GL/C-E7
Ground Equipment	GL/C-E8
Aerial Torpedoes	GL/C-E9
Materials	GL/C-E10

After selecting one or more projects the participating firms were contracted to build a "spatial model", a mockup. The resulting wooden model was then examined by employees of the Erprobungsstelle (E-Stelle, Testing Department), during which instrument operations, armament and spatial data were checked out. Generally, after several changes and improvements, a contract was awarded for the building of one or more test prototypes (Versuchsmuster, or V-Muster). During the manufacturing process as well as throughout the factory testing stage the E-Stelle exercised constant supervision of prototype testing and technical evaluation. Once tentative construction guidelines had been established, details were worked out regarding installation of equipment in the airframe, to include directions for armament and testing equipment. Again, the E-Stelle was heavily involved in this process as well.

After these matters had been satisfactorily sorted out the Technisches Amt reviewed the existing possibilities for producing the proposed aircraft design in series.

In cases where series production was approved, the Chief of the Technisches Amt pronounced the design ready for introduction and decided on the production of a null-series (0-Serie), or pre-production batch. At this point the designated manufacturer was required to provide the relevant equipment listing as well as all design blueprints, plus the anticipated material needs.

Prior to initiating the production of a pre-series, Fachgruppe C II (Fertigung, or Production) examined and tested all inspection gauges to be used, a measure which ensured conformity of subassembly manufacturing. Only then was a final contract awarded. At the same time provisional operating directives were established for the pre-production aircraft. During production the manufacturing process came under the constant scrutiny of specialists from the Technisches Amt. After factory test flights had been completed a few examples were sent to E-Stelle Rechlin and there checked out on an individual basis. Depending on the importance of the aircraft, in some cases the airframes were tested at the E-Staffel level for longer periods of time under combat conditions.

Once this phase had been concluded to the satisfaction of the Technisches Amt, Berlin then reviewed its willingness to procure the initial production series (Baureihe/BR 1). The Chief of the RLM reached the final decision regarding series production (production models usually being designated A-1) and at the same time made the determination who would eventually be given the large-scale production contract.

If the firm previously entrusted with development and the manufacturer assigned production were not one and the same, renewed examination of the remaining bids took place before any license production documents were issued.

Following a contractual agreement and the establishment of service and maintenance directives – which would later evolve into Luftwaffe service directives (LDv. or Luftwaffendienstvorschriften) – the planned full-scale production could finally begin.

While the new aircraft design was still under technical analysis, the Luftwaffe units designated as its planned recipients were given the documents necessary for the upcoming introduction into service, thereby at least giving the units theoretical familiarity with the new model.

Under the guidance of the E-Stelle, so-called field testing got underway after prototype testing had been concluded. The detailed evaluation results found the testing and familiarization reports not only portrayed all the positive aspects of the future production design, but also showed its weaknesses. Efforts were concentrated on full-scale production after the technical changes had been discussed with the manufacturer.

If hidden shortcomings continued to crop up during the course of operations – as was the case with nearly every airframe introduced – Department C II studied the options of improving the aircraft or its equipment with the least amount of outlay in work and material. This measure generally resulted in a number of unnecessary difficulties being avoided even before mass production had begun. Work-intensive modifications on the airframe and engine frequently took place as a result of the often changing mission roles. The changing war situation often caused various designs to undergo radical changes. An example of this was the high-speed Do 335; during its developmental gestation of just two years the prototype underwent many redesigns from its originally intended function:

09/28/1942 Development as a high-speed single-seat bomber.
11/11/1943 High-speed bomber, add'l roles as heavy fighter,
            night fighter and reconnaissance.
01/10/1944 Recon, later as heavy fighter, high-speed bomber
            and night fighter.
03/31/1944 Development as heavy fighter, later planned as
            high-speed bomber.
06/16/1944 Planned exclusively as high-speed bomber.
08/03/1944 Development as night fighter, with possibility of
            high-speed strike role.

09/26/1944 Development as night fighter canceled, now: heavy
fighter.
10/10/1944 Development only as night fighter.

The constant reworking of the design, along with the inclusion of the most diverse engine types, caused significant delays in the work program at the Dornier plant. Thus, the first Do 335 A-6 night fighter prototype was not ready for flight until 15 November 1944. The Reichsluftfahrtministerium proved to be just as indecisive with nearly all high performance aircraft.

Factory Testing and the K.d.E.

The favorable conclusion of factory testing was crucial, particularly that conducted under the jurisdiction of the Kommando der Erprobungsstellen (K.d.E.), for the primary reason that it prevented an otherwise large investment in materials and work effort in poor performing designs. Different E-Stellen were responsible for specific areas, e.g. land- or seaplanes, armament or equipment. On 1 January 1944 the organization was broken down as follows:

Site	Commander	Critical Testing Area(s)
K.d.E.	Oberst Petersen	
E-Stelle Rechlin	Major Daser	General Testing
E-Stelle Travemünde	Major Linke	Seaplanes
E-Stelle Tarnewitz	Major Bohlan	Guns/Cannon
E-Stelle Jesau	Major Stams	Ordnance
E-Stelle Udetfeld	Hptm. Zober	Rocket Armament
E-Stelle Munster-Nord	Stabsing. Dr.Pritzkow	Weapons Testing
E-Stelle Werneuchen	Major i.G.Cerencer	Radio/Radar Equipment
E-Stelle Arktis-Finse	Stabsing. Pantenburg	Winter Equipment
E-Stelle Cazeaux	Hptm. Schlockermann	Ordnance

Aircraft were tested in practice and improved upon at all factory sites, such as at Heinkel in Oranienburg or Junkers in Dessau. This was done regardless of whether the airplane was a result of a direct contract from the RLM or simply an in-house company venture, designed to gain a timeliness advantage for an upcoming contract bid.

For this purpose companies had their own testing centers, such as the Junkers Flugzeug und Motoren Werke (JFM) site in central Germany. Installation of weapons on a trial basis were often carried out there, and new wing and rudder designs were tested in wind tunnels or in flight as part of fundamental research.

The Erprobungsstelle Rechlin (E-St.R.) was located near the Müritzsee and was broken down into seven sub-departments, each of which were under the direction of a department chairman and designated as Erprobungsabteilung E1 to E7. For example, E3 was responsible for all engine development. The director of the E-Stelle, for its part, answered directly to the Technisches Amt, which – as mentioned earlier – was in turn directly subordinate to the RLM.

The immense work expenditure needed to be overcome is illustrated by the official RLM list, which ran from April 1930 to April 1945 and encompassed 635 entries with aircraft designations. Of these, only 298 actually existed, of which just 150 were actually

Comprehensive wind tunnel testing in the late summer of 1941 was directed towards improving the aerodynamics of the He 280's fuselage. For this reason the model was fitted with shortened wings.

The wind tunnel model of the Blohm & Voss P 208 fighter project with the notable AS 413 engine, summer 1944.

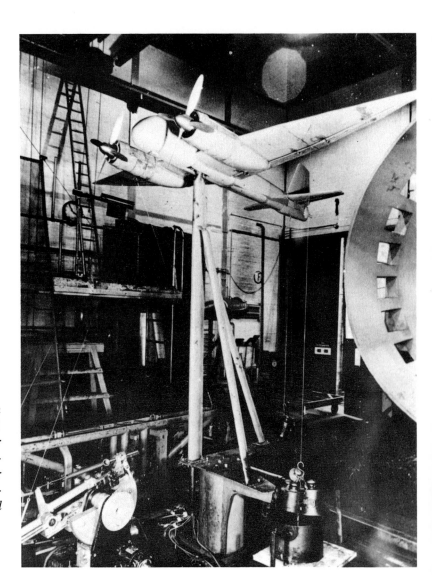

Small-scale wooden model of the Ju 388 K-1 high altitude bomber with twin MG 131 machine guns in the rear gunner's position, photographed during wind tunnel testing.

Various engine configurations were tested using this small-scale copy of the EF 112 (from which the Ju 287 was later derived).

The wooden mockup of the Ju 87 "Stuka" from 1935. It is interesting to note the aerodynamically smooth lines of the oil cooler. This approach, however, was not applicable in practice since the cooling surface was found to be of insufficient dimensions.

As with all subsequent production models, the future Fw 190 also began with an assembly mockup, which was to be fitted with a BMW 139 radial engine rated at 1500 hp. A snow ski rests beneath the 1:1 scale model.

The search for a favorable cockpit arrangement led to – as seen here – the construction of a cockpit mockup. The Ju 85 A represented a high speed bomber project from 1935, powered by two Jumo 211 A or B engines.

The mockup of the Ju 288 B medium bomber with Jumo 222 engines. In September 1941 plans called for a chin turret using an MG 131 Z.

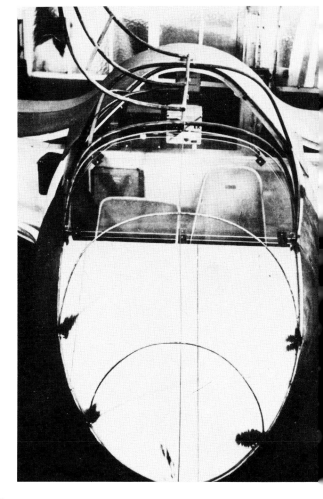

The Lippisch P 10 was to have gone into large-scale production as a high-speed bomber; however, it was not to go beyond the 1:1 scale wooden mockup stage.

Close-up photo of the forward area of a Ju 388 J-1 with SN-2 radar and FuG 212, showing the armament mockup. Series production of this high-performance high-altitude night fighter was never realized due to the war's end.

flown –105 of these in Rechlin. In addition to 16 airframes reserved exclusively for experimentation, 13 of these were unmanned designs, 27 were floatplanes and flying boats, while the remaining 242 aircraft in testing were single- or multi-engine landplanes. Naturally, these figures do not take into consideration that for most of the basic models there were ten, twenty, or even thirty or more sub-variants being evaluated, (for example, a designation brought about when an engine assembly was exchanged for a higher performing engine type). These included efforts to improve cold starting under winter conditions, increasing the service ceiling of operational aircraft, and developing for production and operations the turbine jet engines then in its infancy.

Despite the war situation, K.d.E. testing continued uninterrupted nearly up until the end of the war, although some facilities had to cease operations (e.g. Jesau, Udetfeld, and Cazeaux). In Rechlin, the following were under evaluation at the end of January 1945 as part of the so-called "Führer Emergency Program":

Department:

E2	Do 335/Do 635, Hü 211, Ho 229, Me 262 C-1a/-2b, helicopters
E3	DB 605D, BMW 801 TS, DB 603l and the MeP 8 propeller
E4	Radio emergency program (simplified radio equipment)
E5	Flux gate compass, directional controls, color film, and 3-axis controls
E6	Flare and marking munitions
E7	Bombs (emergency program), TSA 2D, BZA, Lofte 7 and Lofte 8
E10	Factory, replacement and operational materials

In 1945 there were also a number of programs being carried out with several various types of equipment and weapons at the subordinate sites:

Tarnewitz:
Sondergeräte (specialized equipment) SG 113A, SG 117, SG 500, R 100 BS, Panzerblitz Pb 1 and Pb 2, Wgr. 42 air-launched mortar and anti-aircraft gun mounts

Karlshagen:
8-344 and 8-347 air-to-air missiles, 8-246 anti-aircraft targeting device, 8-117 surface-to-air missile, also BMW 003R and Ar 234 with Hs 293 guided bombs.

Travemünde:
Mines, "Kurt" device, Mistel, high-speed drops with L 10, 11, 30 and 40 aerial torpedoes, Fw 190/Ar 234 with torpedoes.

Stade:
Entire radio direction finding program.

One of the last known testing reports is from 4 March 1945 and provides a glimpse of the ongoing activity:
Along with ten Ar 234s, five Do 335s, several Fw 190 D-11/12 and Ta 152s, various Fw 190s were flown with the improved methanol-water booster device. As a result of the more and more infrequent fuel deliveries testing was cut back considerably. The same applied to the deliveries of new prototypes, which could not be supplied on time and waited in a non-flyable state at the manufacturer's, victims of the parts shortage.

Testing of training and operational aircraft which were not fully developed did not proceed without losses. No less than 200 flying crew members met a pilot's death in this endeavor, many of these pilots respected long before the war.

There was also the occasional evaluation of captured enemy airplanes. This was also not without its dangers for the pilot involved, since – although flyable aircraft were frequently captured – in nearly every case the pilot's handbook and operating instructions were missing.

This role was later assumed by the Versuchsverband des Oberbefehlshabers der Luftwaffe (O.b.d.L., Test Unit of the Luftwaffe Chief of Staff)

On August 25th, 1944, the E-Stelle Rechlin was the victim of an Allied bombing raid. This attack resulted in critical evaluation programs being shifted to Lärz. Shortly before the war's end a portion of the remaining programs were transferred to Lech and Memmingen. On 10 April 1945 a second attack took place which nearly brought the activities in Rechlin to a standstill.

The same significance was attached to the entire weapons testing program, which encompassed the MG 15, 17, 81, 131, 151, the MG FF, and various gun mounts and high-caliber cannons. Testing was conducted at Tarnewitz, along with the complex evaluation of air-to-air projectiles. As the war drew to a close, Tarnewitz became the testing site for new types of aircraft armament, such as the "Rohrblock" (SG 117), "Jägerfaust" (SG 500) and new types of reflective gunsights (EZ 42). High-performance Fw 190, Ta 152 and Me 262 aircraft were the most commonly used platforms provided for testing on the ground and in the air.

The evaluation of seaplanes primarily took place at Travemünde. Construction was begun on a site, initially "disguised", as early as 1922; from this sprang the maritime station for the Reichsverbands der Deutschen Luftfahrtindustrie (RDLI, Reich Unit of the German Aviation Industry). In 1928 this became the Seeflugerprobungsstelle Travemünde, which in turn eventually became the E-Stelle (See). There, seaplanes from Blohm & Voss, Dornier, and Heinkel were put through their paces. The final programs included refitting the Do 18 as a submarine killer, testing of droppable lifeboats, the creating of efficient air-sea rescue equipment for single- and multi-engine high-performance fighters and the testing of new kinds of maritime ordnance.

Radio equipment testing took place in Werneuchen. Up until the beginning of 1945, advanced development of radio and radar equipment was carried out in close cooperation with the Funkforschungszentrum (Radio Research Center) in Oberpfaffenhofen.

The Erprobungskommando (EK) was subordinate to the K.d.E. and was assigned the task – as opposed to the Erprobungsstaffel – of using various operational prototypes to evaluate equipment and weapons in realistic situations. Using the results, it explored possibilities for future operations or provided detailed technical requirements.

For example, EK 25 and EK 26 were entrusted with the evaluation of high caliber weapons and mortars. At the same time, experimentation was conducted with towed bombs, the so-called "Grosszerstörer" concept armed with up to 32 air-to-air mortars, and other new types of weapons proposals. The Luftwaffe's E-Stellen in Peenemünde/West and Karlshagen were involved with the testing of glide and guided bombs, missiles, and homing weapons. Particular emphasis was placed on the Fi 103 "Kirschkern", the Hs 293/298, and also the high-speed Me 163 B-1/B-2 rocket fighter.

Aside from the various E-Stellen and Kommandos, other entities were also heavily involved in matters of technical aviation experimentation. These included the Aerodynamische Versuchsanstalt (AVA) Göttingen, the Deutsche Forschungsanstalt für Segelflug (DFS) e.V., and research facilities in Braunschweig, Stuttgart and Berlin.

The Luftwaffe's Erprobungsstaffeln

The Erprobungsstaffeln (E-Staffeln or ESt., Test Squadrons) of the Luftwaffe played a substantial part in the evaluation of combat aircraft under potential operational conditions. Working in close cooperation with the aviation industry, dedicated E-Staffeln, e.g. for the Ju 88, He 177, Ju 188, He 219, Ju 388, Ar 234, Me 262 and the Do 335, were established with command subordination. As late as 5 February 1942 – in spite of the prevailing critical war situation -- the following test and evaluation units were on register:

Versuchsverband OKL
- 1/Versuchsverband OKL with:
    - Einsatzkommando Götz (Ar 234 - Reconnaissance)
    - Einsatzkomando Sommer (Ar 234 - Reconnaissance)
    - Einsatzkommando Braunegg (Me 262 - Reconnaissance)
- 2/Versuchverband OKL (test Staffel for enemy aircraft)
- 3/Versuchsverband OKL (test flight Staffel)

Erprobungskommando 335 (Do 335, Mengen/Württemburg)
Lehr- und Erprobungskommando 16 (Me 163, Brandis)
Erprobungskommando 4 (Fw 190 with X4 missiles)
Erprobungskommando 26 (engaging heavy bombers)
Lehr- und Erprobungskommando V1 (Fi 103, Karlshagen and Nordholz)
Erprobungskommando Kolb (MG 151/20 in triple mounts)
Erprobungskomando 41 (Forestry Protection Staffel)
U-bootjagd-Übungsstaffel (diverse aircraft)
Erprobungskommando 388 (Ju 388 J, Rechlin)
Versuchs- und Einsatzverband Propellermittel der Luftwaffe (diverse aircraft, Berlin-Adlershof)
Verlastungskommando für Transportflugzeuge der Luftwaffe (diverse aircraft)
Ballonstaffel (diverse balloons, Werneuchen)

The following is but a single example of one of the Kommandos: after the first Do 335s had successfully completed their factory and E-Stellen evaluation phases, the Oberkommando der Luftwaffe (OKL) established the "Erprobungskommando Do 335" on 4 September 1944. This organization consisted of a commander, the personnel of a heavy fighter Staffel and the maintenance section detached from the General der Jagdflieger (GdJ, General of Fighter Pilots), and an additional 3 pilots and 37-man technical personnel team assigned from the General der Aufklärungsflieger (GdA, General of Reconnaissance Pilots). The EK 335 was planned as the basic stock for units converting to this aircraft type and was to be combat evaluated as a Mosquito hunter and night fighter, then as a reconnaissance fighter and bomber. The unit maintained this sequence for its testing program up until the time it was disbanded.

On 14 February 1945 the OKL ceased on-going testing operations for EK 335 together with those for Lehr- und Erprobungskommando 16, EK 26, and EK 388 and dispersed the released personnel to several other units, such as JG 400 (Me 163) and SG 151 (Fw 190).

Just prior to the end of the war the remaining Erprobungsstellen and units were broken up. Several of the pilots joined up with the jet fighter units of the IX Fliegerkorps, regrouped on 23 November 1944 and reequipped with jets or single-seat propeller-driven aircraft. Others were transferred to Jagdgeschwader 7 "Nowotny" or integrated into III(Erg.)/JG 2 and its Me 262 A-1a jets.

The Heinkel He 112, here the first edition of the V 3 (Werknummer 1292), was, along with the Bf 109, planned as a replacement for the Ar 68 and He 51 single seat fighters. The Bf 109 won out.

Aside from Bf 109 VB (B-05), two other Messerschmitt fighters also won trophies at the international flight competition in Dübendorf near Zürich on 29 July 1937.

Belly landing of Messerschmitt's Bf 109 A-1, Werknummer 808, on 7 April 1937 in Rechlin. It was caused by a defective crankshaft bearing in the Jumo 210 D engine. Total damage amounted to about 30 percent.

On 18 June 1936 the RLM issued a contract for fitting the Bf 109 with a radial engine. Bf 109 V21 (WerkNr. 1770, D-IFKQ) came from the E-series and later received a 14 cylinder twin radial engine. Dr. Wurster flew this aircraft in August 1939 in Augsburg.

The Focke-Wulf Fw 190 V1 seen in the final stages of assembly. Development of the BMW 139 engine, which was planned as the initial powerplant, was canceled on 30 September 1938.

On 1 June 1939 Flugkapitän H. Sander flew the Fw 190 V1 on its successful maiden flight. In this photo the V1 still carries the markings D-OPZE; later the test aircraft was fitted with a new NACA engine housing and wore the military registration FO+LY. The men in front of the aircraft are (l. to r.): Gen.Ing Lucht, Gen.Oberst Udet and Dipl.Ing. Francke.

Takeoff run measurements with the Messerschmitt Me 262 V3 in Leipheim on 26 March 1943 (flown by company pilot Ostertag). According to the flight report the Jumo T-1 engines performed flawlessly that day. The only problem noted was trouble with exhaust air in the cockpit.

The Fw 187, with twin Jumo 210 Da engines, was a competitor of the Bf 109. The design was rejected, since the RLM considered two engines on a fighter to be purely wasteful – even though the Fw 187 was 60 kmh faster than the Bf 109 B-2. The photograph shows the Fw 187 B1, D-AANA, which was rolled out in Bremen in the spring of 1937.

Final preparations for the record-setting flight of the He 119 (D-AUTE) on 22 November 1937 on the 1000 km route Hamburg-Stolpe-Hamburg. Flugkapitän Hitschke and Dieterle made this record flight. Average speed was 504.988 kmh; maximum speed 620 kmh.

The He 119 made use of a DB 606 double engine with 2700 hp. This was built from two DB 603 engines coupled together. The respectable performance of the aircraft, however, was shortly afterwards overtaken by the Breda 88. A second attempt failed when the machine was forced to make an emergency landing in Travemünde, where it crashed.

In response to a request for a single engine dive bomber Heinkel offered the high-performance He 118 as an option. The first test version (D-IKYM) was still fitted with a Rolls-Royce engine and enlarge cooler.

At the Zürich flight competition the Dornier Do 17 MV1 (D-AELE, WerkNr. 691) won in its class. The aircraft was significantly faster and better performing than any other bomber of its time. The test aircraft continued to fly up until the spring of 1945 with the DFS in Ainring.

The Heinkel He 111 V3, as the final prototype for the bomber series, completed it first flight in April of 1935. The test version for the A-series was initially fitted with two BMW VI engines.

By modifying an He 111 D-0 at the start of 1938 the first prototype for the E-series was born with the V 10. This version was equipped with Jumo 211 A-1 engines. Aside from a new, retractable cooler and modified exhausts, the type was also recognizable by its higher speed of 390 kmh and its greater payload of 2000 kg max.

The Ju 88 V1 completed its maiden flight on 21 December 1936. Werknummer was 4941, designation D-AQEN. The aircraft was powered by two 12 cylinder DB 600 Aa with 1000 hp each.

This early test model of the Ju 88 (VE+AG) found use as a training aircraft for the FFS in Fürth. The second prototype, flying for the first time on 10 April 1937, differentiated itself from its predecessor primarily by an improved engine cooling system.

The He 177 V7, the second armed prototype, was also used from 28 June 1943 as a training aircraft.

Along with serious powerplant difficulties, several landing gear collapses – caused by too high a weight – led to landing accidents with the Junkers Ju 288 V2. The crash shown here occured on 20 July 1942.

Transporting the Dornier Do 335 V1 (WerkNr. 230001) on its way to its maiden flight, which took place on 26 October 1943 by Dipl.Ing. Dieterle in Mengen, Württemburg. The aircraft carried the call letters CP+UA.

Detail view of the DB 603E forward engine in the Do 335 V1, shortly after installation of both engines.

Production

The German Aircraft Manufacturers and Their Most Important Products

The complete list of all aircraft manufacturers and suppliers which initially outfitted the numerous flying clubs and later, the expanding Luftwaffe, is so large that only the most significant companies can be named here. By far the most important were Arado, Dornier, Focke-Wulf, Heinkel, Junkers and Messerschmitt. Additionally, a nearly inconceivable number of supplier firms saw to it that final production assembly never came to a standstill.

Many of the smaller manufacturers were either taken over by the larger companies during the course of the war or, as license builders, strove to ensure that the varying major production programs reached the numbers set forth in the delivery schedules.

Along with the well-known Junkers Flugmotoren Werke some of today's world brand names, such as Daimler-Benz, were even then numbered among the most significant suppliers, particularly in the areas of aircraft electronics or engine design.

This does in no way disguise the fact that the larger companies, too, such as Junkers or Heinkel, held foreign production licenses and – particularly before the war's outbreak – produced countless trainers. Here now is a listing of the most important aircraft manufacturers and their products, produced in both large- and small-series:

AGO-Flugzeugwerke GmbH (Oschersleben bei Magdeburg)	Ao 192, Ar 240, Fw 190, Go 145, He 46, He 51, Hs 123, Bf 109
Albatros-Flugzeugwerke GmbH (Berlin-Johannisthal	L 57 to L 102, Fw 44, Fw 56, Fw 190
ATG-Maschinen GmbH (Mockau bei Leipzig)	Ju W 33, W 34, Ju 86, Ju 88, Ju 188, Ju 388, parts for He 111
Arado Flugzeugwerke GmbH (Warnemünde, Brandburg/Havel)	Ar 65/66/68/76/79/80/81/ 95/96/195/196/197/199/240 /440/232/234, He 177
Bachem-Werke GmbH (Waldsee in Württemburg)	Ba 349 and various license production
Blohm & Voss Flugzeugbau GmbH (Hamburg-Finkenwerden, Wenzendorf)	Ha 135/36/137, BV 138/139 /140/141/142/144/155/222/ 238/, Do 23/17, Ju 86, Me 262
Bücker Flugzeugbau GmbH (Rangsdorf bei Berlin)	Bü 131/133/134/180/181/ 182, Fw 44, DFS 230, Ju 87 parts, Hs glide bombs
Deut. Forschungsanstalt für Segelflug e.V.(Griesheim, Ainring)	various gliders, DFS 194 228/346/230/231, DFS 232

Dornier Werke GmbH (Friedrichhafen, Oberpfaffenhofen, Wismar and various supply facilities)	Do H/J/K/R/S/X, Do 11/13 /23, Do17/217/317, Do 15/ 18/24/26, Do 335, plus license Ju W 33, He 111, Ju 88,, Me 410, Fw 190
Erla-Maschinenwerk GmbH (Mockau bei Leipzig)	Me 5a (Erla 5), license He 51, DFS 230, Bf 109
Fieseler Werke GmbH (Kassel and surrounding areas)	F1 to Fi 97/99, Fi 156/ 256, license AR 68, Fw 58c, He 45, He 51/72, Kl 35, Bf 109, Fw 190
Flettner-Flugzeugbau GmbH (Berlin Johannisthal	Fl 184/185/265/282/285/ 339 helicopters and miniature heli development
Focke-Achgelis & Co. GmbH (Delmenhorst)	Fa 223/330, var. projects
Focke-Wulf Flugzeugbau GmbH (Bremen, Cottbus, Hannover, Posen, Neubrandenburg, Sorau, etc.)	A1-A43, Albatros L102, Fw 44/56/58/189, Fw 190/200, Ta 152/153/154, license Go 145, He 45, He 60
Gothaer Waggonfabrik AG (Gotha)	Go 145/146/149/150/241/ 242/244, license Ar 66, DFS 230, He 45, Bf 110 Bf 109
Heinkel Flugzeugwerke AG (Rostock, Rostock-Marienehe, Oranienburg, Vienna region and various other sites)	HE (Eindecker) and HD (Doppeldecker), He 1, HD 55, He 45/46, He 51, He 59/60, He 70/170/270, He72, He 111, He 100/112/ 114/115, He 177, He 118/ 119/176/178, He 219, He162, license Ju 88
Henschel Flugzeugwerke AG	HS 212, Hs 122/126, Hs 123/129, Hs 128/ 130, guided wpns, license Do17/23, Ju 188
Hellmuth Hirth Versuchsbau GmbH (Stuttgart-Feuerbach)	various experimental designs
Wolf Hirth GmbH - Flugzeugbau (Nabern/Teck)	Gö 8/9, Hütter night fighter, parts for Me 321/323, Bf 109, Ba 349
Horton Flugzeugbau GmbH (Cologne Bonn, Göttingen, Hersfeld)	H I-H XVIII flying wings
Junkers Flugzeug- u. Motorenwerke AG (Dessau, Magdeburg, Leipzig, Fritzlar, Merseburg, Halberstadt, and various other sites)	Ju F 13/G 24/31, W 33/34, Ju 49/EF 65, Ju 52, Ju 60/160, Ju 86, Ju 87, Ju 88 /188/288/388/488, Ju 90/290/390, Ju 287, Ju 248 (Me 263), license He 111
Klemm-Flugzeugbau GmbH (Böblingen)	L 15-l30, Kl 31-152, license Me 162, DFS 230
Messerschmitt Flugzeugwerke AG (Augsburg, Regensburg, Giebelstadt, Leipheim Obertraubling, Schwäbisch-Hall, Wiener Neustadt)	S 15 - M 36, Bf 108, Bf 109, Bf 110/210 /Me 410 Me 209/309, Bf 161/162/ 163, Me 163/263, Me 262, Me 321/323, license U 12, Ar 66, Do 11, Go 145, He 45, He 50

Mitteldeutsche Metallwerke GmbH (Erfurt and surrounding areas)	Ta 154, Ka 430, parts for He 111, He 177, Go 242,Fw 190, Ta 152(planned)
Siebel Flugzeugwerke KG (Halle/Saale)	Fh 104, Si 204, license Fw 44, He 46, Ju 88/188, parts for Do 177
Weser-Flugzeugbau GmbH (Bremen)	We 271, license and parts for Ju 52, Ju 86, Ju 87, Fw 200, Ta 154
Luftschiffbau Zeppelin GmbH (Friedrichshafen)	follow-on development of Me 323 and "Rammer" miniature rocket fighter

The authorization of each one of these aircraft prototypes was contingent upon a corresponding "Dringlichkeitsstufe", or priority class, based upon the progress of the war or the expectations of the OKL. On 7 September 1943 the proposal for the necessary priority classification of all general types of aircraft was as follows:

Role	Priority Class	Model (in order of priority)
Fighter and Nightfighter	1	Fw 190, Bf 109, Bf 110, He 219, Ju 88 C-6, Me 262, Me 163, and Ta 154
Heavy Fighter (Zerstörer)	2	Me 410, Bf 110, and Hs 129
Vengeance Weapons (Vergeltungswaffen)	3	Fi 103 "Kirschkern" (V1)
Bomber	4	Ju 88/188, He 177, Ju 87, and Ju 288
Long-range Recce	5	Ju 290
Trainers	6	Ar 96/396, Bf 108, Bü 181, and Si 204
Transports/ Air-Sea Rescue	7	Ju 52/352, Do 24, and Me 323
Seaplanes	8	Ar 196
Liaison	9	Fi 156
Cargo Gliders	10	Go 242 and DFS 230

The production figures which the German aviation industry were to realize, despite widespread air strikes and grievous material bottlenecks, are shown in a chart produced by the Generalquartiermeister, 6 Abteilung, of 2 March 1945. This not only shows the detailed program expectations, but also includes the number of operational aircraft actually produced by the manufacturers.

Day fighters	1944									1945	
Type	June	July	Aug.	Sep.	Oct.	Nov.	Dec.	Jan.	Feb.	Mar.	Apr.
Bf 109 quota:	1150	1230	1530	1774	1850	1230	1426	210	1850	1820	–
actual:	1120	1237	1138	1511	1505	1312	016	1208	948	–	–
Fw 190 quota:	640	740	930	1045	1285	1100	1265	882	1470	1325	–
actual:	554	752	880	885	673	997	918	882	544	–	–
Me 163 quota:	4	11	28	50	92	50	50	–	–	–	–
actual:	3	12	13	35	61	22	90	30	3	–	–
Me 262 quota:	–	–	5	35	50	175	195	167	511	722	–
actual:	–	–	5	19	52	101	124	10	280	–	–
Ta 152 quota:	–	–	–	–	10	10	20	20	45	145	–
actual:	–	–	–	–	–	–	18	20	3	–	–
He 162 quota:	–	–	–	–	–	–	–	50	100	202	–
actual:	–	–	–	–	–	–	–	–	82	–	–

Night fighters

Type	1944							1945			
	June	July	Aug.	Sep.	Oct.	Nov.	Dec.	Jan.	Feb.	Mar.	Apr.
Do 335 quota:	–	–	–	–	–	–	–	–	12	5	–
actual:	–	–	–	–	–	–	–	–	–	–	–
Others quota:	385	430	790	467	414	365	405	370	–	–	–
actual:	397	417	439	490	353	426	230	251	107	–	–
Heavy fighters											
quota:	146	125	80	43	3	8	3	5	–	–	–
actual:	109	93	68	40	3	1	3	1	4	–	–
Strike fighters											
quota:	505	660	612	517	400	300	350	367	350	350	–
actual:	507	623	563	574	413	294	312	367	309	–	–
Bombers											
Ar 234 quota:	–	5	10	20	40	40	35	35	28	10	–
actual:	–	5	10	18	40	40	35	35	15	–	–
Me 262 quota:	40	60	95	82	175	–	–	–	–	–	–
actual:	28	59	15	72	65	–	–	–	–	–	–
Ju 388 quota:	–	–	–	–	–	–	–	–	–	–	–
actual:	–	–	–	–	–	1	1	2	–	–	–
other quota:	284	313	294	56	–	–	–	–	–	–	–
actual:	231	329	165	29	–	2	2	–	–	–	–
Reconnaissance											
tact. quota:	150	225	150	150	100	177	174	–	–	–	–
actual:	110	111	237	94	78	151	70	2	63	–	–
strat. quota:	69	87	69	71	54	47	78	–	–	–	–
actual:	65	85	35	66	51	61	73	47	21	–	–
Transports											
quota:	65	64	50	–	–	–	–	–	–	–	–
actual:	59	84	23	2	1	–	–	–	–	–	–
Seaplanes											
quota:	18	14	18	–	–	–	–	–	–	–	–
actual:	19	15	13	3	3	–	–	–	–	–	–
Trainers											
quota:	308	342	393	275	260	262	251	170	97	–	–
actual:	331	332	391	319	271	274	252	–	–	–	–

In order to make life difficult for enemy intelligence and at the express wishes of Hitler, on 26 March 1945 cover names were established for all war-critical aircraft, armored vehicles and special weapons. As far as it affected the Luftwaffe, the ten most important models were assigned official cover names:

Ta 152	fighter, reconnaissance and heavy fighter (Butcher-bird)	"Würger"
He 162	fighter (Sparrow)	"Spatz"
Me 163	fighter (Comet)	"Komet"
Me 262	fighter, reconnaissance and fighter-bomber (Storm-bird)	"Sturmvogel"
He 219	night fighter (Eagle Owl)	"Uhu"
Do 335	night fighter and heavy fighter (Arrow)	"Pfeil"
Ju 388	night fighter and long-range reconnaissance (14th cen. German pirate)	"Stortebecker"
Me 410	heavy fighter and reconnaissance (Hornet)	"Hornisse"
Ar 234	reconnaissance and bomber (Lightning)	"Blitz"
Ju 188	long-range reconnaissance and torpedo bomber (Avenger)	"Rächer"

Nevertheless, designations – even if they were falsely applied after the war - have been preserved down even to today in reference material. Examples of this are "Kraftei" (Powered Egg) in place of "Komet" for the Me 163, or "Hecht" (Pike) for the Ar 234. Other designation, such as that for the Ju 188, were seldom used and today are no longer seen.

Dispersal of Production

During the course of the war the Allied air superiority became more and more oppressive. Starting in 1942, powerful armadas of enemy bombers not only became the terror of the civilian populace, but also increasingly struck against the German armament industry. All efforts to effectively protect important sites for any length of time by utilizing fighters or anti-aircraft guns were doomed to failure against formations of 800-1000 attacking bombers. Even more so, it was absolutely futile to protect the cities; on average, one person was killed for every three tons of bombs dropped.

The only chance to protect at least the industrial operations against such destruction was to scatter them into outlying areas, which for the time being would be outside the bombers' range. The Rüstungskommando accordingly arranged it so that entire factories would be shifted to Silesia (Arado), to Poland (Heinkel) or to Czechoslovakia

(Messerschmitt). The air attacks began in earnest in the summer of 1943 and led to a production loss of more than ten percent in the first half of the year. The US 8th Air Force and to a lesser extent, the Royal Air Force, bore the brunt of these attacks. Were it not for preparations begun in 1941 to disperse a portion of production, the resulting massive jump in production in 1944 would never have been possible. For example, the number of Focke-Wulf factories at its subsidiaries increased from five on 31 December 1943 to twelve in 1944. Thanks to the excellent transportation system, which, although damaged, had not yet been hit hard, it was possible to bring the components produced at scattered sites together for final assembly.

At the same time "dummy factories" were built, which mirrored their real counterparts in both size and shape and served to distract the enemy bomber crews. An example of this was a site established near Ploesti in Rumania which was bombed by the Soviet air forces for years before finally being recognized as a mockup near the end of the war. The RAF was also similarly fooled, bombing a dummy Krupp-Werk near Essen at the beginning of 1943. Yet despite all the concentration of flak emplacements around critical facilities the surest protection remained in "bunkering" the factory or setting up underground production centers. In Germany there were 140 larger underground factories in April of 1945 alone, in which nearly half of all aircraft engines, a high percentage of ammunition and nearly all "vengeance weapons" were produced. Particularly large sites were situated at Leonberg near Stuttgart, in Nordhausen, at Kahla in the Harz, and also at Igling near Landsberg and Mühldorf in Lower Bavaria. There production of the Me 262 ran quite smoothly, safely beneath the cover of a meter-thick concrete canopy. Work was also undertaken on a second huge bunker in Kaufering, which by the spring of 1945 had reached enormous dimensions. It was planned to use that site for the production of jet fighters beginning in the fall of 1945. The end of the war brought Plan "Kuckuck" (Cuckoo) to an abrupt halt.

Just as important was the production of fuels and lubricants to cover the needs of the armed forces. Luftwaffe units in particular required sufficient amounts of fuel to fulfill their task of defending the Reich homeland. For the processing of fuels additional hydrogenation plants were set up in Leuna, Boehlen bei Leipzig, Meerbeck, Lützken-dorf bei Halle, and near Hannover. As with production of engines and airframes, much of these sites remained empty shells. For during the interim the Allied bombers had begun to destroy the German infrastructure. Accordingly, beginning in 1944, road and rail lines, as well as waterways and canals became targets of both the USAAF and the RAF. More and more frequently supply transports were destroyed before reaching their intended destinations. Tank cars burned in the stations and on the railroad lines following strafing attacks. Thus, the activities of the Luftwaffe were forcibly crippled.

Aircraft Equipment and Armament

The Luftwaffe and armaments leadership attempted to beat the enemy's superiority with quality – not quantity. The goal was the so-called "high performance airplanes," which were to be superior to all enemy types in terms of ceiling, climb speed and maximum speeds.

But even as early as 1943 there were already deviations from this concept. On the 18th of November, Luftwaffe headquarters became involved with the new operational designs, which utilized turbojet or rocket engines. The offensive potential was still at the forefront: "Jet bombers are to be pushed into operations in large numbers with all resources available, using small bombs as defense against an invasion in the West; jet bombers are more important than jet fighters. At that time work was only underway on the single-seat non-pressurized Me 262 jet fighter, the Me 163B rocket fighter, and the Ar 234 jet reconnaissance aircraft. All three types were still in the testing phase, however. The jet bomber had not even reached the conceptual phase.

Despite the developmental stage of the above-named high-performance aircraft, the use of the Me 163 and Me 262 was seen as an effective means of countering high-altitude enemy bombers. However, equipping the first units with these new fighters was not expected to take place until the spring and summer of 1944. In spite of the requirement to produce the Me 262 as a fighter-bomber (as it was offered by the company), the GdJ and the OKL pressed for immediate acceleration of series production of the fighter. At the same time it became possible to fit the aircraft with easily-produced conversion kits, turning it into a high-speed tactical bomber able to carry up to 500 kg of bombs.

Despite all efforts to the contrary, it was not possible to set up the first jet bomber Staffeln prior to the Normandy invasion. More and more frequently the Allied forces were able to seize the initiative. The battered German divisions retreated from France, fighting all the way. At the same time the air war over the Reich intensified. For these reasons, it became necessary to increase fighter production using every means conceivable. A secret brief from 30 June 1944 gave the following directions:

Production lines for the He 111, He 177, Ju 290 , Ju 352 and the trusted Ju 52 were to cease immediately. All released workers were to be assigned in up to three shifts for daily production of fighters for the protection of the Reich. The powerplants needed for piston-engine fighters and strike aircraft could be achieved –thanks to sufficient stocks on hand of DB 605 and BMW 801 engines – through remanufacturing operations.

At the same time orders were given to massively escalate new production. The Ju 88 night fighter was to be fitted with the Jumo 213 A inline engine in place of the BMW 801 radial, as was the Fw 190 D-9 and its successors, the Ta 152 E and H.

By the end of 1944 Me 262 production had finally gotten underway in earnest. The production of twin-engine medium range bombers had, for the most part, already come to a standstill, since – other than in the East – these types were no longer capable of operating with any type of real success. Reequipping of Jagdgeschwader (JG) 7 and Kampfgeschwader (KG) 51 was initiated.

In addition to the turbojet aircraft, the K.d.E. anticipated the delivery of new piston-driven designs to beef up the fighter units. Highlight of these efforts were the Bf 109 K-4 and the Fw 190 D-9, plus the Ta 152 H, a high-altitude fighter with an impressive wingspan and DB 605 engine. The Bf 109 H had already been canceled and the BV 155 was not really considered, since the first prototype could not be expected before the beginning of 1945.

With regards to bombers, the Ju 388 K-1 with Jumo 213E or BMW 801 TI was to be temporarily supplemented with the Do 335. The Ar 234 C became the focal point of interest as a four-engine high-speed bomber, while the Ar 234 B-1 and B-2 were seen only as an interim stop-gap solution. As a heavy bomber, the Reichsluftfahrtministerium planned for the series production of the Ju 287 six-engine jet bomber. These intentions had not yet been fully implemented, since the BMW 003 turbojet initially operated quite unreliably and there was not yet any effective, remote controlled gun system available.

The Luftwaffe leadership envisioned the Ju 88 G-7 in use alongside the improved-performance G-6 along with the reduced-weight He 219 with Jumo 213 inline engines, which was to operate against attacking Mosquito bombers flying at ever higher altitudes. In this vein, the desire was more and more frequently expressed to fit the aircraft with the high-performance DB 603 piston engine as soon as possible.

Aside from the Bf 109 as the standard trainer, there were also the Do 335 A-12, the Me 262 B-1a, and the Ta 152 "S" to be employed as conversion trainers. The production of sufficient quantities of aircraft engines was fundamental to ensuring that production goals were at least generally adhered to. The following detailed chart provides a clear-cut view of the quantities produced as well as the production statistics during the critical months of December 1944 and January 1945:

Engine type	Emergency power at maximum pressure altitude	Installed in aircraft type	Monthly production (quota)	New production (actual) Dec. 1944	New production (actual) Jan. 1945	Prior total deliveries	Stock on hand	Reserve stockpile
Jumo 211 F/J/K/P	F=1190/5.0 J=1330/4.6	Ju 88 Ju 87 He 111 H-6	–	–	–	50908	1706	3749
Jumo 211 A/B/D G/H/L	945/4.4 1190/5.5	Ju 87 B,R He 111 H	–	–	–	17068	939	1065
Jumo 213/8013	1500/6.0	Ju 88 Bf 109	1521	982	829	7229	188	2
Jumo 004 TL		Me 262	1400	801	864	4169	13	–
DM 603	1620/5.7	He 219 Do 335	253	407	237	8753	270	1218
DB 605	1355/5.7	Bf 109 Bf 110	2000	1865	1430	41099	1834	919
DB 606	2640/4.8	He 177	–	–	–	494	–	–
DB 610	2710/5.7	He 177	–	–	–	2656	52	21
BMW 132 A/T/Z K/N	Z=550/2.0 N=690/1.8	Ar 196 Ju 52	–	–	–	23564	931	2940
	1500/4.6	Fw 190	–	–	–	3893	8	–
BMW 801 A/L BMW 801 C/D	C=1500/4.6 D=1560/5.7	Fw 190	300	470	345	18129	218	–
	1560/5.7	Fw 190	703	526	500	6684	333	–
		Ar 234	150	94	93	267	–	–
BMW 8801 BMW 003 TL	850/4.2	Fw 200 Do 24	–	–	–	4088	254	1286
BMW 323	400/3.0	Ar 96	150	150	50	7705	475	355
As 410 As 411	516/3.0	Ar 396 Si 204	250	104	250	4642	187	925
As 10 C/P	200	Fi 156	130	90	130	22080	1274	730
HM 500	105	Bü 181						

Altogether an exact total of 223,434 aircraft engines were produced up until the end of January 1945 and, with the exception of the so-called OKL reserves, nearly all were installed in operational and training aircraft. The following list of engines manufactured at the BMW-Werke is provided as an example of the broad spectrum of production engines, not all of which necessarily entered series production and a few of which were used only as experimental designs.

Type	Series	Year	Description	hp/rpm
BMW 139	-	1938	14-cy. air-cooled twin radial fuel-injected engine	1200 hp 1900rpm
Fafnir 323 A,B	2	1935	9-cyl. air-cooled four-stroke fuel-injected radial (high-altitude)	660 hp 2100 rpm
Fafnir 232 C,D	2	1935	9-cyl. air-cooled four-stroke fuel-injected radial engine	670 hp 2100 rpm
Fafnir 232 Q	2	1938	9-cyl. air-cooled fuel-injected radial engine	660 hp 2100 rpm
BMW 801 A	1 2	1940	14-cyl. air-cooled twin radial with two-stage supercharger, complete assembly with flush bearing ring and engine casing, design tropical-resistant but not -proof	1600 hp 2100 rpm
BMW 801 B	-	1941	Engine not produced	
BMW 801 C	-	1941	14-cyl. air-cooled twin radial fuel-injected, "bare" engine with propellor pitch control gear	1600 hp 2100 rpm
BMW 801 D	-	1942	14-cyl. air-cooled twin radial fuel-injected, "bare' engine with propellor pitch control gear and auxiliary starter	1600 hp 2100 rpm
BMW 801 E	-	1943	14-cyl. air-cooled twin radial fuel-injected, improved performance BMW 801 with modified casing (experimental only)	1800 hp-2500(max) 2200 rpm
BMW 801 F	-	1944	14-cyl. air-cooled twin radial fuel-injected, improved BMW 801 without housing (experimental only)	1800 hp 2200 rpm
BMW 801 G	-	1944	14-cyl. air-cooled twin radial fuel-injected with BMW 801 bomber engine	1600 hp 2100 rpm
BMW 801 H	-	1944	14-cyl. air-cooled twin radial fuel-injected with BMW 801 fighter engine	1600 hp 2100 rpm
BMW 801 J	0 1	1944	14-cyl. air-cooled twin radial fuel-injected, on roller and glide bearings, turbine drive (high-altitude engine)	1600 hp 2100 rpm
BMW 801 L	-	1944	14-cyl. air-cooled twin radial fuel-injected with mounting kit (series production planned)	1600 hp 2100 rpm
BMW 802	-	1940	18-cyl. air-cooled twin radial fuel-injected engine (experimental only)	2800 hp 2300 rpm
BMW 803	-	1943	28-cyl. air-cooled four-bank radial, fuel-injected engine (experimental only)	3000 hp 4100 rpm

Manufacturing the BMW 801D at the main plant in Munich-Allach. In order to facilitate greater mobility of the aircraft engines they were mounted on roller carriages.

The average weight of the BMW 801A twin radial engine as installed in the Fw 190 A was 1200 kg, with a rating of 1400 hp.

Checking over an Fw 190 A-3 with a BMW 81 D-2 engine. In order to perform the necessary work on the landing gear the aircraft was raised up on jacks.

The indispensable final inspection of the Fw 190 A-6. Production aircraft were fitted with two MG 17 and four MG 151/20 guns.

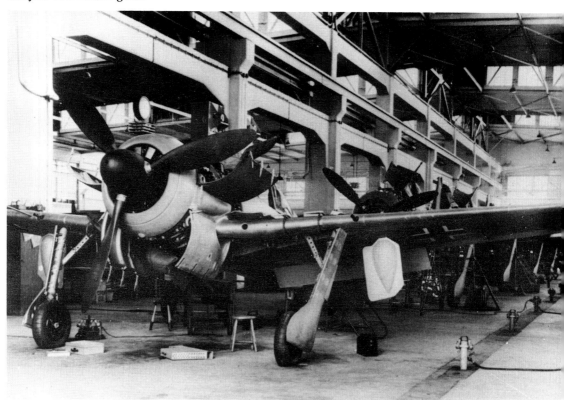

Weserflug in Bremen played a key role in the production of the Ju 87.

Final assembly stage of the Ju 87 B included the installation of the Jumo 211 engine.

The assembly line of the legendary Ju 87 "Stuka" in the massive hall of the Berlin-Tempelhof airport.

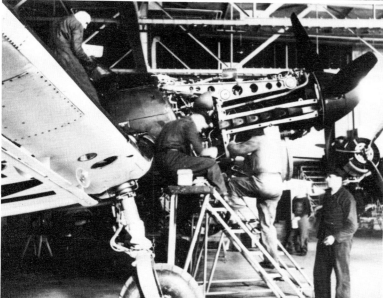

Final factory checks of a Ju 87 B-1 manufactured in Dessau.

Damaged Ju 87 B fuselage portions following a devastating bombing raid on Plant 2 of Weserflug GmbH.

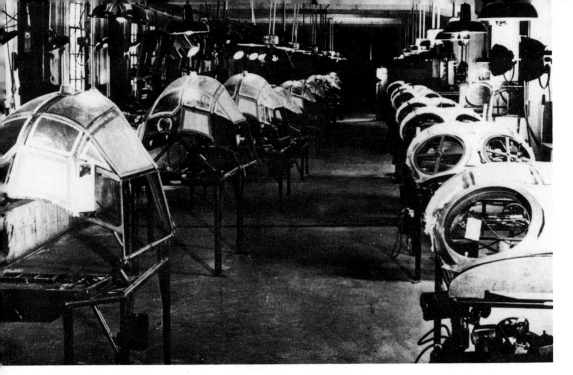

Assembly of the canopy glazing for the Junkers Ju 88 series at Opel in Rüsselsheim.

Series production of Ju 88 wings at Halberstadt, Central Germany.

In Aschersleben the fuselages for Ju 88 A-1s through A-4s were manufactured on the assembly line.

Final assembly of the Ju 88 A-5 being carried out in Bernburg.

Engine mechanics at the Junkers-Motorenwerke here put the final touches on the Ju-88's Jumo 211 engine.

These two photos and the one on the following page show the assembly of the Heinkel He 111 H-6 in the autumn of 1942. The photographs were probably taken in the modern facilities of the Oranienburg plant.

Drop Ordnance and Aerial Torpedoes

At the beginning of the war the Luftwaffe had only a limited selection of ordnance available to it, most of which was introduced back in 1933. This included the 1-kg electron incendiary bomb, the SD 10 fragmentation bomb, plus two general purpose bombs, the SC 50 and the SC 250. In 1935 the SC 500 was added, at the time the heaviest general purpose bomb. One year later the two GRUNDMINEN, the LMA and the LMB (500 and 1000 kg, respectively), were introduced. In 1938 the 1000 kg F 5 aerial torpedo followed, as did the SD 50 general purpose bomb. One year later the 1.3 kg B1BE electron incendiary bomb with steel nose, the SD 500 general purpose bomb and, also weighing 500 kg, the PC 500 anti-armor bomb were all added to the inventory.

The battles in Poland and Norway had already clearly demonstrated the lack of adequate ordnance types. In particular, relatively few outstanding successes were recorded against armored warships with the existing types of bombs. This deficiency led to the introduction, beginning in 1940, of the SC 1000, SC 1800, and SD 1700 general purpose bombs. At the same time the bomber and dive bomber units of the Luftwaffe were given the PC 1400, which permitted attacks on heavily protected targets from both horizontal flight as well as a diving profile. Attack units received the SD 2, the SBe 50 (fragmentation bomb/concrete weighing 50 kg) and two liquid incendiary bombs, the Flam C 250 and 500.

Along with the anti-personnel incendiary bomb (Strbd C 500) and a more powerful electron-incendiary bomb, the B2E(Z) weighing 2 kg, the SD 250 general purpose bomb appeared with increasing frequency beginning in 1942, as did the BM 1000 aerial mine and the heavy SC 2500 general purpose bomb. Finally, there was the "heavy load" bomb, the SB 2500. This weapon was only used for a short time and had a length of approximately 3.90 meters. A smaller version of the SB 2500, on the other hand, was often carried by the Do 217 E-4 and K-1, as well as the He 177 A-3 and A-5. Later, the remaining SB 2500s were filled with the so-called Trialen-Mischung 105 for successful strikes against merchant shipping, even with near misses by up to 70 meters. The explosive effect of this bomb when detonated on land was so great that considerable damage occurred within a radius as far out as 1000 meters. Starting in 1942 – by which time the fighting had intensified in the Soviet Union – a smaller fragmentation bomb, the SD 1 was supplied to the Luftwaffe for the upcoming battles, as was the 4 kg SD 4 HL hollow charge bomblet. The SD 70 was also developed for the close air support role. The SD 1000 and the remotely guided PC 1400X, also called the "Fritz X" or "FX", were employed against well protected targets. In the following year of the war the ordnance was tailored to tactical needs even more so. For attacks against city centers the light liquid-filled incendiary bomb, the Brand 10 was introduced along with the high-intensity incendiary Brand C 50 and C 250 bombs. In addition to the SD 65, the SB 1000 heavy load bomb was also developed and supplied to the front units, as was the SC 2000 and two armor piercing bombs, the PD 500 and 1000. However, the PC 500 RS, PC 1000 RS and PC 1800 RS propelled armor piercing bombs were only produced in limited numbers and were eventually brushed aside in favor of guided ordnance like the

Hs 293 A-1/A-2 or the PC 1400X. The so-called drop containers, designated AB, began to make their appearance no later than 1944. Representative types were the AB 70, AB 250, AB 500 and AB 1000. These offensive weapons, e.g. the Ab 1000, could be filled with a maximum of 610 B1 incendiary bombs, 238 Brand 2, 78 Brand 10, 66 Brand 10 and 60 Brand 1, or with 204 SD 4 bomblets. The heavy drop canisters were used quite frequently against London during Operation "Steinbock." The AB 70 was used in large numbers within Luftflotte 6 and the West right up until the end of the war.

By mid-March the heavy Type 1400 torpedo bomb had joined various other aerial torpedoes undergoing testing in Travemünde. Testing of special ordnance, such as the "Winterballon", could not be carried out due to the predominant weather conditions in the spring of 1945 and the critical fuel situation.

In addition, at that time there were also numerous bombs filled with chemicals; aside from the KB 3 and KB 10 there were also large stockpiles of KC 250, 500, and 1000 chemical bombs in various depots. Massive quantities of these special bombs were haphazardly buried during the closing weeks of the war and even today still pose a potential threat which should not be underestimated.

Barrel and Rocket Armament

Within the concept of offensive and defensive aircraft armament, the equipment of Luftwaffe's operational units initially appeared quite modest. The light MG 15 was the primary moveable gun for bombers, fitted with a 75 round magazine, whereas the MG 17 was primarily installed in a fixed mounting. The first of several heavy automatic weapons was represented by the MG FF (caliber 20 mm).

Prior to the beginning of the war, trials also began with the MG 81 I and Z as well as the MG 131, with series production already gearing up. Both weapons later became the standard weapon for nearly every multi-engine bomber in the Luftwaffe's inventory. The MG 81 Z was first installed in the He 111's side gunner positions. Aside from various fighter aircraft, the MG 81 found widespread use in the Do 17, Ju 88, He 111 and later the Bf 210/Me 410 and the He 177.

During the course of the war's first two years the MG FF was replaced more and more frequently by the two versions of the MG 151. This was a 15 mm weapon, rapidly supplemented by a 20 mm version and most commonly used as a fixed gun in both single- and multi-engine fighters. Additionally, weapons gondolas were developed, designated as Waffentropfen (WT) or weapons teardrops, and often seen on the Bf 109 G-6. The MG 151/20 was installed as an engine cannon beginning in 1944, utilized in the Fw 190 D-9 and D-11 and the Ta 152 H-10 high-altitude fighter introduced in 1945. The cannon was primarily used as an offensive and defensive weapon in the A-Stand (nose compartment) of multi-engine bombers such as the Fw 200 C-3 or the He 177 A-5. Beginning in the fall of 1943, thorough experimentation was done on new types of gunner's position using twin mounts. Multi-barrel weapons were not given priority for series production, since on the one hand the heavy weight of the weapon wasn't considered satisfactory and on the other hand the manufacturers simply didn't have the capability to produce them. And finally, there was the

3 mm Maschinenkanone (MK) 108, which was the standard weapon for all Me 262 A-1a jet fighters and Me 262 A-2a Blitzbombers (BB) from 1944 on.

Although in much lesser numbers, the MK 108 was also utilized in the form of a Rüstsatz, or conversion kit, on the Bf 109 K-4 –and the Fw 190 – as a gondola-mounted weapon. There was also consideration given to fitting the He 162 Volksjäger with the cannon. However, since the weapon was prioritized for installation in the Me 262, it was not felt that there would be adequate stock; this decision in turn led to virtually all He 162s being equipped with the MG 151/20 from March 1945.

Other than the weapons briefly outlined above, the Luftwaffe also made use of several cannon types developed for engaging ground targets, some even being installed during series production. One of these weapons, the MK 101, was conceived back in 1936, but prototype testing dragged out over several years. Along with installation in the Bf 110 and production models of the Hs 129, two of these 30 mm cannons were installed in a few He 177 A-1s, which were then given the designation "Zerstörer" (Destroyer). It was planned to use these planes to intercept four-engine Allied bombers far out into the north Atlantic during their transition flights and shoot them down prior to reaching England. After testing by the K.d.E., however, the idea was scrapped, apparently because an OKL directive specified that attacking the enemy's rail traffic was more important.

At least one He 177 was also fitted with an MK 103 in the A-Stand. Aside from that, this heavy gun was also installed on production models of the Me 410 A-2/U4, also being planned as armament for the Do 335 B-2 and the Hs 129 B-3.

The Flak 18 gun was tested as heavy armament under the wings of the future Ju 87 G-1 for engaging enemy tanks. As a result, the G-2 was produced in series; the 37 mm gun was also installed below the fuselage of the armored Hs 129 B-2 attack plane. Finally, Ju 88 Ps underwent testing in Tarnewitz with two BK 3.7 guns, during which one of the aircraft suffered a collapsed landing gear.

The Bordkanone 7.5 belonged to the heaviest class of aircraft weapons; it was derived from the 75 mm Panzerabwehrkanone (KwK) 40 L and was fitted beneath approximately 24 aircraft into a large fuselage "tub", at the opposite end of which an MG 81 Z was located for defensive weaponry. Whether there were actually several He 177s with the BK 7.5 continues to be a mystery. Due to numerous technical problems with the ammunition feed and removal on the Hs 129 B-3, there were endless delays during the testing of this flying tank killer. At least two prototypes were undergoing trials in the fall of 1944, but development was eventually broken off. It is unknown whether the weapons already produced were ever used in operational aircraft before the war's end.

Aside from fitting barreled weapons, by the end of 1944 several types of air-to-air rockets were in advanced stages of development. In particular, the spin-stabilized R4M rocket was used with great success by both III/JG 7 and Jagdverband 44 (Galland). The wire-and radio-guided versions of the X-4, on the other hand, were only experimental designs; it was already far too late for combat operations. By the end of the war the X-7 "Rotkäppchen" (Little Red Riding Hood) anti-tank rocket was under development, as was the Hs 298 air-to-air missile. Test launchings from a Ju 88 S-1 and a Do 217 N-2 proved successful, but no further advances were made in the program.

The most noteworthy examples of offensive missile systems are the PC 1400 X and the Hs 293 A. These were successfully launched by crews from Kampfgeschwader 40 and 100 over the Mediterranean, the Atlantic and the English Channel and were guided using radio signals into or at least into the vicinity of the targets. However, maritime storms and improper storage of the weapons caused a surprisingly high number of failures. Because of the interference sometimes observed in the radio transmission frequencies, starting in 1944 Hs 293 bombs were equipped with wire-guidance and delivered to at least one Gruppe of KG 100. Due to the shortage of fuel these could no longer be operated from Denmark. In the fall of 1944 the He 177 A-5s stationed there had their engines removed and carted off to be used in the production of fighters.

Further development of the Hs 293 A-1/A-2 were the previously mentioned Hs 293 B-1 and the underwater running C-series missiles. Using a television camera for guidance control, a few of these were field tested off the coast of Pomerania. The TV 143 guided missile, LT 950 torpedo glider, and the LT 350 circling aerial torpedo (of Italian design) didn't always meet the demanding requirements of field testing, either. The same applied to the BV 246 glide bomb, manufactured in several sub-variants and eventually relegated to the role of target for anti-aircraft missile units in training.

Despite the immense progress in development, this advantage was not exploited effectively, particularly when in 1941 numerous projects were considered to be unnecessary and work was broken off for more than a year. To the RLM's officiating "flyers" of the First World War the Bf 109, with its initially notable speed advantage, seemed to be completely adequate.

Defense against low-flying aircraft was provided at nearly all German airfields by the Flak 38 (20 mm).

The 20 mm Vierlingsflak proved substantially more effective against low-flying aircraft – at airfields as well. The photograph shows a fixed site in Hamburg.

The Flak 3.7 cm also found use for protection in nearly every theater of the war.

Engine maintenance for an He 111 H-6. Aircraft of this type – like the Ju 88 – were notable for their easy access to the engines.

Mounting a Jumo 211 to a Ju 88 A-5 of 1/KG 76 by mechanics at a Frontwerft-Abteilung, or Frontal Maintenance Unit.

The engine of this Ju 52 from KGzbV 172 is covered for protection against dust and weather. The crews in the Geschwader were initially recruited at the beginning of the war from the employees of "Deutsche Lufthansa."

This Ju 88 A-5 from KG 54 shows its dive brakes to good effect.

Engine change for an He 177 A-5 using a special type of crane which could be mounted to the wings and greatly simplified the process of switching out the (initially) problematic engines.

A field hoist tripod is used to replace a BMW 801 radial engine after it had been damaged in a crash landing.

Inspecting a 2000 hp DB 605 AS in a Bf 109 K-4 during the last winter of the war.

After loosening the four speed bolts the forward lower landing gear shroud could be removed without difficulty.

The hydraulic brake lines and leather cover for protecting the solid inner strut can be seen here. The braked wheel was 815 x 290 mm.

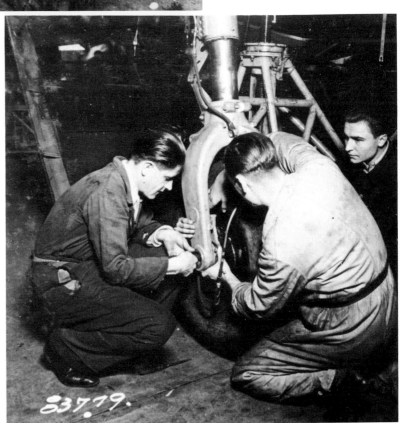

Working on the forked strut of a Ju 87 B-1.

The Do 217 K-1, here with a midnight black underside for night operations over England, had double braked wheels 1200 x 450 mm.

The landing gear of the He 111 with 1140 x 410 braked wheels, was retracted hydraulically and extended by means of a spring tension reservoir. Landing gear movement was accomplished electrically or, in case of emergency, mechanically.

The production Einhängerost 4 (ER 4), which held four SC 50 bombs. The aircraft in this photo is a Bf 109 F-4/R-6.

Two SC 50 bombs under the left wing of a Ju 87. This payload (cylindrical fragmentation bombs) had a total weight of 50 kg, of which 25 kg was explosive material. The bomb had a length of 1095 m and a diameter of 200 mm.

Loading an He 111 B-0 with SC 250 bombs. These were carried vertically in the bomb bay. Lifting the bombs was accomplished using a rope pulley.

Detail shots of the attachment point for the SC 250 underneath the fuselage of a Ju 87 B. Note the trapeze assembly, which was designed to ensure the bomb cleared the propeller arc during a diving attack.

It will probably never be known whether this message ("Happy Easter, anno 1942") was meant in earnest.

Four SC 250 bombs carried as external ordnance under the inboard wings of a Ju 88 A-6 converted back into an A-5. The weight of the explosive charge in each of these bombs was 135 kg.

Two SC 500 bombs carried in the bay of a Do 217 E-4 from KG 2 "Holzhammer." The photo was taken in the winter of 1942/1943.

Transporting a heavy SC 500 using a loading trolley to a Do 217 E-2 of 5/KG 6 (3E+EN) parked near a camouflaged hangar. For operations as a night bomber the aircraft carries a rear radar warning receiver.

Interesting writing on an SC 1000 suspended beneath the fuselage of an He 111 H.

The SC 1000 "Hermann" had a length of 2.80 m and a diameter of 0.65 m. With an explosive charge of 530 kg, the bomb had a destructive radius of 350 m. It could make a crater up to 10 meters deep.

In addition to HE bombs, incendiary bombs were also utilized quite frequently, carried in AB 1000 drop containers. After being dropped, 610 small B-1 bomblets were released and dispensed over an area of approximately 2000 square meters.

Weighing in at 2430 kg, the SC 2500 "Max" was the heaviest bomb in the German Luftwaffe. The explosive charge accounted for 1700 kg of the total weight.

Special weapons testing using an Fw 190. Notice the camera mounted beneath the wing of the single-seat fighter.

An invention of H.G. Bätcher is revealed in this photo. Using two containers of flammable liquid welded together, the nose and tail of an SC 1000 are then attached. The four small bombs on the side act as the primers. The bomb is then dropped into a river and, following the explosion, the entire area is turned into a lake of fire. The bomb was to be used to set fire to wooden bridges.

Sighting in and checking the firing pattern of the two MG 17s which fire through the propeller arc of the Bf 109 B.

Adjusting the wing guns of a Ju 87 B-1. The flag serves as a safety measure and shows the state of firing readiness.

Armorers checking and cleaning the MG 17 guns of a Bf 110 E-2 of 5 Staffel (Pik-As) of II/ZG 26.

Detail photo showing the two MG 17s in an Fw 190 A-1.
A Bf 109 G-6/R-6 with MG 151/20 wing gondolas during the test flight. The MG 151/20, cal. 20.0 mm, was derived from the MG 151, cal. 15.1 mm.

An Me 410 A-4 Zerstörer (Destroyer) with MG 151 cannons, which were retrofitted after manufacture. Previously the aircraft was armed with a single BK 5 and bore the designation Me 410 A-4/U-4.

Last-minute checks for the test version of the MK 214, installed in the Me 262 A-1a/U-4 (V083) "Pulkzerstörer" (Formation Buster) for firing trials at the end of February 1945.

The designers of the Ju 288 anticipated the use of two fixed MG 151/20 cannons, seen here in a mockup photo.

Production design of the Ju 388 J night fighter with two BK 3.7 cm and FuG 212 search radar, from 1944.

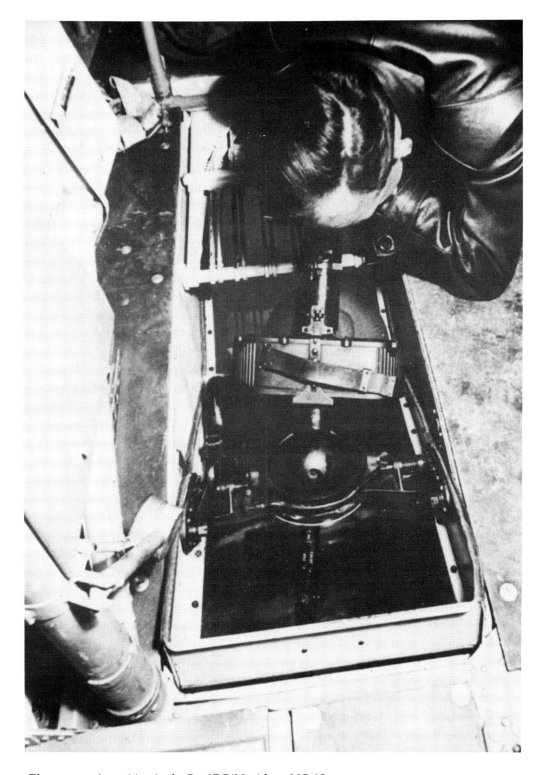

Floor gunner's position in the Do 17 P/M with an MG 15.

Interior shot of the forward gunner's position (A-Stand) in an He 111 H-3 with the MG 15. The MG 15 was developed from the MG 30 in 1932 by Rheinmetall-Borsig. The field of fire was approximately 35 x 50 cm at a range of 100 m.

The defensive armament on an He 111 H-11 included the MG 17, which was of the same caliber as the MG 15 but had a higher rate of fire (1200 rounds per minute [r.p.m.] compared to 1050 r.p.m.). The penetrating force was approximately 5 mm of armor plate.

The MG 81 was a follow-on development of the MG 15 with a rate of fire of 1600 r.p.m. Here the gun is mounted in the side station of an He 111 configured as an MG 81 Z (Zwilling, or twin).

Remote-controlled defensive armament on the fuselage side of an Me 410 A-4.

The Ju 88 V27 (WerkNr. 7027, D-AWLN) was a prototype for the future Ju 188 series. First flight of the V27 took place on 27 December 1941.

The 1:1 scale mockup of the planned Junkers Ju 188 R with FuG 202 and two MG 151 in the upper forward gunner's compartment (B-Stand). The rear-firing defensive position was armed with a single MG 131.

Mockup of the nose compartment of a Ju 290 A-7 with MG 151/20 Zwilling and an FuG 200 surface search radar. The test flights of this type began in the spring of 1944 and production began in the summer of the same year.

Manufacturer's photograph of the HD 151 weapons system, most commonly used in multi-engined bombers.

Ventral compartment with sight – but minus weapon – of the Ju 290 A-3 (WerkNr. 0157), converted from the A-1. The aircraft began initial testing on 19 June 1943 at the E-stelle, Rechlin.

Side view of the MG 151, which had a length of 1917 mm. In the background can be seen a Ju 88 prototype with partially recognizable callsign + SO, in Dessau.

The design for the MG 131 was born in 1933. Testing began on 1 July 1938 in, among others, the fuselage position of the Dornier "Wal."

Modified rear gunner's compartment mockup of the Ju 290 with a single MG 151/20.

Wooden model of the HL 131Z/1 rear gunner's position on a Ju 188, which was tested in November of 1943 in Tarnewitz. Later testing of the production version was considerably delayed due to poor weather.

Realization of the previously seen wooden mockup of the HL 131Z/1 on a Ju 188 G, here the Ju 188 V1, which first flew on 12 June 1943. The planned Ju 188 G-2 version was to have been powered by twin Jumo 213 A-2 engines.

Model of the HL 131V (V = vier, or quad) rear gun assembly, which in 1944 was planned for use on the Ju 290.

Test installation of the HL 131V at the rear of an He 177 in 1943.

Remotely-operated MG 131 rear gun of a Ju 388 during testing in Dessau.

Production prototype for the Ju 88 P-1 with BK 7.5. The length of the weapon was 6105 mm, weighing 705 kg. The BK 7.5 was developed from the PaK 40L anti-tank gun.

Field testing of the Ju 88 P-1. Each of the high-explosive armor shells had a weight of 11.9 kg; twelve were carried in the magazine.

The Ju 87 D-4, planned as a torpedo bomber for the LT F 5B torpedo, was based on the D-1 and D-3 series. The type never saw service since both the He 111 J-6 and Ju 88 A-17 were superior in terms of range, speed and carrying capacity.

Beginning at the start of 1942 the F 5B torpedo was being used operationally. The aerial torpedo had a length of 4.98 m and a diameter of 0.45 m. The photo shows a training version, clearly recognizable from its paint scheme.

*The F 5B had a 200 kg warhead and reached a speed of 60 kmh in the water. Like the Ju 88 A-4/torp.,
the He 111 H-6 could carry two of the torpedoes.*

A live LT F 5B under the fuselage of an He 111 H-6, (VE +CH), in the late summer of 1943.

In order to stay out of the range of ships' anti-aircraft defenses, the LT 950 was fitted with a glider. This permitted the torpedo to be released at a safe distance. The drawback was, however, that the ship's crew usually had enough time to maneuver out of harm's way after observing a drop. Notice the prominent guidance section at the tip of the upper portion of the glider.

A concrete torpedo (LT 950 B) during the testing of the glider in 1942.

Practice LT 950 torpedo in front of a Ju 188 A-3 in late summer, 1944. The callsign of this medium bomber was PL+TS.

The Luftwaffe in Action

Units and Organization of the Luftwaffe

On 8 May 1920, the German Luftstreitkräfte was officially disbanded by the Chef der Heeresleitung, General von Seeckt. In Versailles, the victorious powers had agreed upon a German Heer encompassing only 100,000 men with strictly defined equipment.

Only after 14 April 1922 could Germany again build aircraft, but even then only without weapons and of limited performance. The 100 Fokker D XIII single-seat fighters ordered from the Netherlands during the Ruhr Crisis are not known to have become operational, but they did later serve to form the basis of the Lipetsk Training Center in the Soviet Union. Working in close cooperation with the Red Army, beginning in 1924 pilots were not only given operational training there, but the evaluation of new equipment was also undertaken there under operational conditions. Furthermore, by 1 November 1930 approximately 300 Army and 40 Navy pilots had been trained at the site as a safety precaution against the so-called A-Fall (Case "A", the outbreak of war). Using diverse, meaningful camouflaged titles, from then on the secret training became even more intensive. At the same time, Deutsche Lufthansa was drawn into the training program. The so-called "Reichsbahn rail lines" served to provide future bomber crews familiarization in night and poor weather flying.

By September 30th, 1935, a goal of 4021 aircraft were to be acquired. At the end of 1934 676 had already been delivered, plus an additional 1238 trainers. The first Aufklärungsstaffeln were established at the beginning of 1935, equipped with the He 45/46. The speedy He 70 arrived a short time later, and in 1937 it was joined by the Do 17, followed by the Ar 68, the He 51, and finally the (for its time) exceptionally fast Bf 109 fighter.

The smallest flying unit in the Luftwaffe throughout this time was the Staffel, consisting of 12-15 aircraft; three or more of these Staffeln formed a Gruppe, three of which in turn formed a Geschwader. Several Geschwader comprised a Fliegerdivision; then came a Fliegerkorps and finally a Luftflotte. The tactical callsigns introduced by the Luftwaffe revealed the subordination of a given aircraft. The following are the known codes:

A1+	Kampfgeschwader KG 53	C2+	Aufklärungsgruppe 41
A2+	I/Zerstörergeschwader ZG 52, later II/Zerstörergeschwader ZG 2	C6+	Aufklärungsgruppe 121 Transportgruppe zbV 600
		C8+	Transportgeschwader TG 8
A3+	Kampfgeschwader KG 200	C9+	Nachtjagdgeschwader NJG 5
A5+	I/ Stukageschwader StG 1, later Schlachtgeschwader SG 1	D1+	Seeaufklärungsgruppe 126
		D5+	Nachtjagdgeschwader NJG 3
A6+	Aufklärungsgruppe 120	D7+	Wekusta 1
B3+	Kampfgeschwader KG 54	D9+	Nachtjagdgruppe NJGr 10 and NJ-Staffel Finland/ Norway
B4+	Nachtjagdgeschwader NJG 3 (Norway)		

E6+	Aufklärungsgruppe 122	S7+	Stukageschwader StG 3, later Schlachtgeschwader SG 3
E7+	Kommando der Erprobungsstell		
F1+	Kampfgeschwader KG 76	S9+	Erprobungsgruppe 210, later Schnellkampfgeschwader 210
F2+	Ergänzungs-Fernaufklärungsgruppe		
F6+	Aufklärungsgruppe 122		
F7+	Seeaufklärungsgruppe 130	T1+	Aufklärungsgruppe 10
F8+	Kampfgeschwader KG 55	T3+	Bordfliegergruppe 196
G2+	Aufklärungsgruppe 124	T5+	1 and 2/Aufklärungsgruppe ObdL, later 1/ Fernaufklärungsgruppe 100
G6+	Kampfgeschwader zbV 6		
G9+	Nachtjagdgeschwader NJG 1, temporary Zerstörergeschwader ZG 1	T6+	Aufklärungsgruppe 122
		T9+	Versuchsverband Obdl
H1+	Aufklärungsgruppe 10	U5+	Kampfgeschwader KG 2
H4+	Luftlandegeschwader LLG 1	U8+	I/Zerstörergeschwader ZG 26
H8+	Aufklärungsgruppe 33	V4+	Kampfgeschwader KG 1
J2+	Nahaufklärungsgruppe 33	V7+	Aufklärungsgruppe 32
J4+	Transportstaffel 290, later Transportstaffel 5	V8+	Nachtschlachtgruppe NSG 1
J9+	I(Stuka)/Trägergeschwader 186	W1- W6+	reserved for Me 331 units
K1+	Stabsschwarm Luftflotte 6	W7+	Nachtjagdgeschwader NJG 100
K6+	Küstenfliegergruppe 406		
K7+	Nachtaufklärungsgruppe ObdL	W8- W9+	reserved for Me 321 units
L1+	Lehrgeschwader LG 1	X4+	Lufttransportstaffel 222, also Seeaufklärungsgruppe 129
L2+	Lehrgeschwader LG 2		
L5+	Kampfgruppe zbV 172		
M2+	Küstenfliegergruppe 106	Z6+	Kampfgeschwader KG 66
M7+	Küstenfliegergruppe 806	1B+	Wetterstaffel 5 (Luftwaffe 5)
M8+	I and II/ Zerstörergeschwader ZG 76	1G+	Kampfgeschwader KG 27
		1H+	Kampfgeschwader KG 27
N3+	Kampfgruppe zbV 172	1K+	Nachtschlachtgruppe NSG 4
P1+	Kampfgeschwader KG 60 (planned)		
		1R+	Kurierstaffel Finland
P2+	Aufklärungsgruppe 21	1T+	Kampfgruppe zbV 126
P4+	X Fliegerkorps command flight	1Z+	Kampfgruppe zbV 1, later Transportgeschwader TG 1
P5+	Sonderstaffel Transozean (Do 260	2F+	Kampfgeschwader KG 54 (transitional)
R4+	Nachtjagdgeschwader NJG 2	2H+	Zerstörerversuchsstaffel 210
		2J+	Zerstörergeschwader ZG 1
S1+	Stukageschwader StG 2	2N+	Zerstörergeschwader ZG 76
S2+	Stukageschwader StG 77, later Schlachtgeschwader SG 77	2P+	X Fliegerdivision command flight
		2S+	Zerstörergeschwader ZG 2
S3+	Transportgeschwader TG 30	2Z+	Nachtjagdgeschwader NJG 6
S4+	Küstenfliegergruppe 506	3C+	Nachtjagdgeschwader NJG 4

Code	Unit	Code	Unit
3E+	Kampfgeschwader KG 6		Stukageschwader StG 1
3J+	Nachtjagdgeschwader NJG 3	6I+	Küstenfliegergruppe 706
3K+	Minensuchgruppe der Luftwaffe	6K+	Aufklärungsgruppe 41
3M+	I/Zerstörergeschwader ZG 2	6M+	Aufklärungsgruppe 11, later Nahaufklärungsgruppe 8
3U+	II/Zerstörergeschwader ZG 26	6N+	Kampfgruppe KGr 100, later Kampfgeschwader KG 100
3W+	Nachtjagdgeschwader NJG 11		
3X+	II/Kampfgeschwader KG 1	6R+	Seeaufklärungsgruppe 127
3Z+	Kampfgeschwader KG 77	6U+	Zerstörergeschwader ZG 1
4A+	IV/Zerstörergeschwader ZG 26	6W+	Bordfliegergrupper 196, later Seeaufklärungsgruppe 128
4C+	Nachtjagdstaffel/KG 40 and Kommando Kunkel		
4D+	Kampfgeschwader KG 30	6Z+	Transportgruppe Herzog
4E+	Aufklärungsgruppe 13	7A+	3/Aufklärungsgruppe 121
4N+	Aufklärungsgruppe 22	7J+	Nachtjagdgeschwader NJG 102
4R+	components of Nachtjagdgeschwader NJG 2	7R+	Seeaufklärungsgruppe 125
4U+	Aufklärungsgruppe 123	7T+	Kampfgruppe KGr 606
4V+	Kampfgruppe zbV 172, later Transportgeschwader TG 4	7Ü+	Kampfgruppe zbV 108
		7V+	Kampfgruppe zbV 700
5D+	Aufklärungsgruppe 31	8H+	Aufklärungsgruppe 33
5F+	Aufklärungsgruppe 14	8L+	Küstenfliegergruppe 906
5J+	Kampfgeschwader KG 4	8T+	Kampfgruppe zbV 800, later Transportgeschwader TG 2
5K+	Kampfgeschwader KG 3		
5M+	Wekusta 26	8V+	Nachtjagdgeschwader NJG 200
5T+	Kampfgeschwader KG 101	9K+	Kampfgeschwader KG 51
6G+	III/Stukageschwader StG 51, later II/	9P+	Kampfgruppe zbV 9
		9V+	Fernaufklärungsgruppe FAG 5
		9W+	Nachtjagdgeschwader NJG 101

From Blitzkrieg to the Defensive

Since so much has already been written about the overzealous buildup of the Luftwaffe and its accompanying mistakes and successes, it is probably sufficient to simply look at a few of the more significant events in a closer light.

As a result of the "Blitzkrieg" successes in Poland and Western Europe, the Luftwaffe High Command initially drew some false conclusions. Close air support, as important as it may be, and a fanatical "Stuka Mania" formed the central points for misinterpretation in later events.

As early as 1935 the book "Air Dominance" (Il dominio dell'aria) by Giulio Douhet had appeared in Germany, on whose themes it would have been easy to focus. His most important points were:

- to achieve air dominance is the same as to achieve victory.
- to be beaten in the air is the same as being hopelessly conquered.
- the air armada must be employed in massive numbers.
- the enemy must be destroyed at his own level, in his production centers.

The friendly heavy bomber forces which Douhet called for, well protected by "air forces of special character" (meaning long-range fighters, according to the original publisher), were to play a decisive role in any future air war. Yet in Germany, even before the war had started, the promising Do 19 and the robust Ju 89 were both canceled, while the medium bomber and a number of varying attack and close support aircraft increasingly found favor. The desire for dive bombers prevented Göring and his influential colleagues from properly interpreting the signs of the times. These individuals became obsessed with the idea that not only the Ju 87, but also all medium and heavy bombers – such as the Ju 288 and the He 177 – were to be unconditionally designed to be capable of dive bombing. The time loss caused by this thinking in planning, construction, and testing was to play a significant factor in influencing the final outcome of the Second World War. For the coming Blitzkriegs against a less well equipped or poorly led army, Luftwaffe units operating Bf 109s and He 111s seemed to fit the bill. For strategic missions, however, a true heavy bomber was lacking, as is clearly illustrated by the Luftwaffe Air Order of Battle from 1 September 1939:

Luftwaffen Lehrdivision

Stab(K)/LG 1	Greifswald	10 He 111
II(K)/LG 1	Neubrandenburg	41 He 111
III(K)/LG 1	Greifswald	40 He 111
10(See)/LG 1	Travemünde	9 He 111
IV(Stuka)/LG 1	Barth	39 Ju 87
Stab/LG 2	Jüterbog-Damm	3 Bf 109
I(J)/LG 2	Garz	36 Bf 109
I(Z)/LG 2	Barth	32 Bf 110 and 3 Do 17
II(N)/LG 2	Garz	10 Bf 109
7(F)/LG 2	Jüterbog-Damm	12 Do 17
8(F)/LG 2	Jüterbog-Damm	12 Do 17
9(H)/LG 2	Jüterbog-Damm	11 Hs 126

Luftflotte 1

1(H)/10	Neuhausen	11 Hs 126
2(H)/10	Neuhausen	12 Hs 126
1(H)/11	Grossenhain	9 Hs 126 and 3 Hs 46
1(H)/21	Stargard	12 Hs 126
2(H)/21	Stargard	12 Hs 126
3(H)/21	Stargard	11 Hs 126
4(H)/21	Stargard	9 Hs 45

1(H)/41	Reichenberg	12 Hs 126
2(H)/41	Reichenberg	11 Hs 126
3(H)/41	Reichenberg	9 Hs 126 and 2 He 46
3(F)/10	Neuhausen	12 Do 17
3(F)/11	Grossenhain	12 Do 17
3(F)/11	Grossenhain	10 Do 17
4(F)/11	Grossenhain	12 Do 17
1(F)/120	Neuhausen	13 Do 17
1(F)/121	Prenzlau	11 Do 17
2(F)/121	Prenzlau	10 Do 17
3(F)/121	Prenzlau	12 Do 17
4(F)/121	Prenzlau	11 Do 17
I/JG 1	Seerappin	54 Bf 109
I/JG 21	Jesau	29 Bf 109
I/JG 2	Döberitz	42 Bf 109
10(N)/JG 2	Fürstenwalde	9 Bf 109
1 and 2/JG 20	Fürstenwalde	21 Bf 109
Stab/JG 3	Bernburg	3 Bf 109
I/JG 3	Zerbst	48 Bf 109
I/ZG 1	Jüterbog-Damm	32 Bf 110
II/ZG 1(JGr 101)	Fürstenwalde	36 Bf 109
I/ZG 2(JGr 102)	Bernburg	44 Bf 109
Stab/KG 1	Neubrandenburg	7 He 111
I/KG 152(later		
II/KG 1)	Neubrandenburg	37 He 111
I/KG 1	Kolberg	38 He 111
Stab/KG 2	Cottbus	11 Do 17
I/KG 2	Liegnitz	37 Do 17
II/KG 2	Liegnitz	35 Do 17
Stab/KG 3	Elbing	11 Do 17
II/Kg 3	Heiligenbeil	36 Do 17
III/KG 3	Heiligenbeil	39 Do 17
I/KG 25	Rechlin	18 Ju 88
Stab/KG 4	Erfurt	6 He 111
I/KG 4	Gotha	31 He 111
II/KG 4	Erfurt	32 He 111
III/KG 4	Nordhausen	33 He 111
I/StG 1	Insterburg	35 Ju 87 and 3 Do 17
I/StG 2	Cottbus	38 Ju 87
III/StG 2	Stolp-Reitz	38 Ju 87
III/StG 2	Langensalza	40 Ju 87
II(Schl)/LG 2	Tutow	40 Hs 123

Luftflotte 2

1(H)/12	Münster-Loddenheide	12 Hs 126
2(H)/12	Münster-Loddenheide	12 Hs 126
3(H)/12	Münster Loddenheide	9 Hs 126, 10 He 46, 3 he 45
4(H)/22	Kassel-Rothwesten	12 Hs 126
1(F)/22	Kassel-Rothwesten	11 Do 17
2(F)/22	Kassel-Rothwesten	12 Do 17
3(F)/22	Kassel-Rothwesten	12 Do 17
1(F)/122	Goslar	6 Do 17
2(F)/122	Goslar	12 Do 17
3(F)/122	Goslar	12 Do 17
I/JG 26	Cologne-Ostheim	48 Bf 109
II/JG 26	Düsseldorf	48 Bf 109
10(N)/Jg 26	Düsseldorf	9 Bf 109
I/ZG 26	Dortmund	52 Bf 109
II/ZG 26	Werl	48 Bf 109
III/ZG 26(JGr 126)	Lippstadt	49 Bf 109
Stab/KG 26	Lüneburg	8 He 111
I/KG 26	Lübeck-Blankensee	32 He 111
II/KG 26	Lüneburg	35 He 111
Stab/KG 27	Hannover-Langenhagen	6 He 111
I/KG 27	Hannover-Langenhagen	34 He 111
II/KG 27	Wunstorf	26 He 111
III/KG 27	Delmenhorst	28 He 111
II/KG 28	Gütersloh	35 He 111

Luftflotte 3

1(H)/13	Göppingen	12 Hs 126
2(H)/13	Göppingen	11 Hs 126
3(H)/13	Göppingen	12 Hs 126
4(H)/13	Göppingen	9 Hs 126 and 3 He 46
5(H)/13	Göppingen	9 Hs 126 and 3 He 45
1(H)/23	Eschwege	12 Hs 126
2(H)/23	Eschwege	12 He 46
4(H)/23	Eschwege	9 He 46 and 3 He 45
1(F)/123	Würzburg	12 Do 17
2(F)/123	Würzburg	12 Do 17
3(F)/123	Würzburg	13 Do 17
I/JG 51	Bad Aibling	47 Bf 109
I/JG 52	Böblingen	39 Bf 109
I/JG 53	Wiesbaden-Erbenheim	51 Bf 109

II/JG 53	Mannheim-Sandhofen	43 Bf 109
1 and 2/JG 70	Nuremberg	24 Bf 109
1/JG 71	Friedrichshafen	15 Bf 109
2/JG 71	Friedrichshafen	24 Bf 109
10(N)/JG 72	Mannheim-Sandhofen	16 Ar 68
I/ZG 26	Dortmund	52 Bf 109
I/ZG 52(JGr 152)	Illesheim	44 Bf 109
Stab/KG 5 1	Landsberg	6 He 111 and 3 Do 17
I/KG 51	Landsberg	36 He 111
III/KG 51	Memmingen	36 He 111
Stab/KG 54	Fritzlar	9 He 111
I/KG 54	Fritzlar	36 He 111
Stab/Kg 55	Giessen	9 He 111
I/KG 55	Langendiebach	33 He 111
II/KG 55	Giessen	31 He 111
III/StG 51	Wertheim	40 Ju 87 and 3 Do 17

Luftflotte 4

1(H)/14	Köttingsbrunn	9 Hs 126 Nd 3 He 45
2(H)/14	Köttingsbrunn	2 Hs 126
3(H)/14	Köttingsbrunn	9 Hs 126 and 3 He 46
1(H)/31	Brieg	9 Hs 126
2(H)/31	Brieg	8 He 46
4(H)/31	Brieg	9 He 46 and 3 He 45
3(F)/31	Köttingsbrunn	11 Do 17
1(F)/31	Brieg	12 Do 17
1(F)/124	Wiener Neustadt	11 Do 17
I/JG 76	Vienna-Aspern	49 Bf 109
I/JG 77	Breslau	50 Bf 109
II/JG 77	Pilsen	50 Bf 109
I/ZG 76	Olmütz	31 Bf 110
II/ZG 76	Gablingen	40 Bf 109
Stab/KG 76	Wiener Neustadt	9 Do 17
I/KG 76	Wiener Neustadt	36 Do 17
III/KG 76	Wels	39 Do 17
Stab/KG 77	Prague/Kbely	37 Do 17
I/KG 77	Prague/Kbely	37 Do 17
II/KG 77	Brünn	39 Do 17
III/KG 77	Olmütz	34 Do 17
Stab/StG 77	Breslau-Schöngarten	3 Ju 87
I/StG 77	Brieg	40 Ju 87
II/StG 77	Breslau-Schöngarten	42 Ju 87
I/StG 76	Graz	39 Ju 87 ad 3 Do 17
Stab/KGzbV 1	Fürstenwalde, Burg	
I/KGzbV 1	Burg	

II/KGzbV 1	Stendal	
III/KGzbV 1	Berlin-Tempelhof	
IV/KGzbV 1	Braunschweig	
Stab/KGzbV 2	Neuruppin	total of 496 Ju 52
I/KGzbV 2	Tutow	
II/KGzbV 2	Fassberg	
III/KGzbV 2	Lechfeld	
I/KGzbV 172	Berlin-Tempelhof	
II/KGzbV 172	Berlin-Tempelhof	
10/KGzbV 172	Berlin-Tempelhof	
/KGzbV 9	Berlin-Tempelhof	56 Ju 52, Fw 200, 1 Ju 90 and 1 Ju G 38

Marineflieger (maritime) Units (NorthSea)

Stab/KüFlGr 106	Norderney	
1(Mz)/KüFlGr 106	Norderney	
2(F)/KüFlGr 106	Norderney	
3(Mz)/KüFlGr 106	Borkum	total of 31 He 59
1(Mz)/KüFlGr 306	Norderney	54 He 60
Stab/KüFlGr 406	List/Sylt	36 Do 18
1(M)/KüFlGr 406	List/Sylt	8 He 115
2(F)/KüFlGr 406	List/Sylt	
3(Mz)/KüFlGr 406	List/Sylt	
I/Bordfliegergruppe 196	Wilhelmshaven	6 Ar 196

Marineflieger (Baltic)

Stab/KüFlGr 500 (KGr 806)	Dievenow	
1(M)/KüFlGr 506	Dievenow	
2(F)/KüFlGr 506	Dievenow	
3(Mz)/KüFlGr 506	Dievenow	total of 27 He 60
Stab/KüFlGr 706	Kamp bei Kolberg	27 Do 18
1(M)/KüFlGr 706	Kamp bei Kolberg	21 He 111
3(Mz)/KüFlGr 706	Kamp bei Kolberg	
2(F)/KüFlGr 606	Kamp bei Kolberg	
5/Bordfliegergruppe 196	Kiel-Holtenau	6 Ar 196
4/S/TrGr 186	Kiel-Holtenau	12 Ju 87
5 and 6(J)/TrGr 186	Kiel-Holtenau	24 Bf 109

Heinkel He 51 B fighters of 1 Staffel, I/JG 132 in September 1936 parked in their hangar. In the background a few civil aircraft can also be seen.

A photo of the Heinkel He 112 V4, prototype for the A-series, during the 1937 International Meet in Dübendorf, Switzerland. The aircraft was also shown at the gigantic aviation exhibition in October 1937 in Milan.

Three He 112 E fighters destined for export during engine warm-up. This model was a modified version of the B-1 series. It was equipped with a Jumo 210 Ea engine rated at 720 hp.

An He 112 B of IV/JG 132 from Oschatz Airfield in September 1938. The aircraft had a wingspan of 9.10 m and a length of 9.30 m. Maximum speed was 510 kmh.

This Bf 109 E belonged to I/JG 136, later to become II/JG 333. After yet another change it became II Gruppe of JG 77.

A trio of hunters.

These two pilots make themselves comfortable in front of their combat aircraft for a typical propaganda photo.

At a Rumanian airfield near Ploesti, these pilots are simulating how to ward off an attack by the USAAF 15th Air Force. A Bf 109 G-2 along with a handfull of Rumanian fighters can be seen in the background.

An unusual method for extracting one's self from a threatening "checkmate" situation, as demonstrated by these pilots from III/JG 3 rushing to their Fw 190 A-1s for a scramble alert.

A pilot on airfield defense duty climbs into the cockpit of his waiting Fw 190 A-3, based at the Cognac airfield in 1943.

An Fw 190 A-2 of JG 5 in Norway.

An Fw 190 of 4/II/JG 54 at an airfield in Finland. A Finnish Brewster "Buffalo" can be seen in the background.

By making use of turbine-engine fighters – such as this Me 262 A-1a in Lechfeld in the late summer of 1944 – the Luftwaffe leadership hoped to force a rapid change in the air war.

Oberleutnant Müller of III(Erg.)/JG 2 on the wing of his Me 262 A-1a, which came off the Leipheim assembly line.

On the 3rd of January 1945, after the installation of the FuG 218 V and FuG 226, WerkNr. 1700056 was transformed into the new, second prototype for the single-seat night fighter; it was also used for braking and stabilization testing.

An Me 262 A-1a of Jagdverband 44 discovered by American troops near Innsbruck.

With the Me 163 B rocket-powered fighter, the fighter command leadership attempted to intercept approaching high-altitude enemy bombers. Due to the limited range of the interceptor, this was not always successful. This photograph shows an aircraft during combat trials with JG 400 in Brandis.

An Me 163 B ready for takeoff, also in Brandis near the end of 1944. The white cloud coming from the fuselage underside stems from the steam-powered starter, which sets the fuel feed pumps into motion.

A rocket fighter of JG 400 taking off. A Mach cone is noticeable at the exhaust opening, caused by the high discharge velocity of the exhaust gasses.

Initially planned as a "Volksjäger", or People's Fighter, the flight characteristics of the He 162 proved to be much more complicated than thought and could only be safely mastered by experienced pilots.

Parked He 162 A-2 fighters of I(Einsatz)/JG 1 in Leck.

Carting off an He 162 A-2 of JG 1 by the Allied troops. The previously introduced unit coat of arms is seen on the fuselage side.

The Bf 110 C-4/B series was equipped with the ETC 250. The aircraft shown here belonged to ZG 1. A characteristic feature of 1 and 2 Staffel was the colorful wasp on the fuselage nose.

Several Bf 110 F-2s in tight formation. Notice that the aircraft in the foreground still has its access ladder extended.

The Bf 110 E-2 was put to use near the end of the war with Ergänzungsgruppen, using fresh pilots retrained in the tactical role.

This Bf 110 G-2 was taken over for training from a Zerstörergeschwader in the Mediterranean theater.

The so-called "Blitzkrieg" against Poland was not the Luftwaffe's "stroll in the park" against a markedly inferior enemy as it is generally portrayed. Evidence from the report by the Generalquartiermeister dated 5 October 1939 reveals that 285 aircraft were total losses and an additional 279 planes were more or less severely damaged. Polish operational losses amounted to 333 of 397 aircraft from units participating in the operations. It was not possible to destroy the Polish forces on the ground beforehand; the "Bomber Brigade" continued until 16 September 1939, flying desperate missions against the advancing German Divisionen. This does not detract from the fact that a directed coordination of close air support and other Luftwaffe operations the advance of the armored and infantry units moved rapidly ahead. The sole heavy air attack struck Warsaw on the 25th of September, 1939, and quickly led to the city's surrender. The experiences gained during the Polish campaign formed the theoretical basis for future operations of the Luftwaffe leadership; the simple word "Blitzkrieg" soon became a much-feared term.

The German strike against France followed a short time later, while at the same time airborne troops attacked important fortified positions and bridges in the enemy's rear areas. As these units didn't have heavy weapons it was difficult to secure the seized area for any length of time. The first massed operations in Norway also proved to be a warning for the Dutch General Staff; we need only to look at the airborne landings in the area of Ypenburg near Den Haag, which resulted in the loss of a total of 220 transport aircraft, primarily Ju 52s. A far greater number of the robust, three-engine transports were out of action for a time after being damaged by flak or "heavy landings." An additional 109 aircraft, of which were 29 Bf 109s and 28 He 111s alone, were also lost. An invasion of the British Isles from the air didn't fail with the end of the Battle of Britain, but as early as back in May of 1940.

On the other hand, the German Luftwaffe succeeded in providing Panzergruppe von Kleist and the 4th and 6th Armee with effective close air support, thereby considerably smoothing the advance of these large units. Thanks to well-trained attack forces and the legendary Stuka, the Ju 87, the Luftwaffe was able to ward off dangerous flanking attacks such as the one on the 21st of May, 1940, near Amiens.

The massed employment of single-engine fighters and the Bf 110 Zerstörers ensured that air superiority was secured over the battlefield while at the same time causing notable losses in the defending ranks of the courageous French Armee de l'Air and the British Expeditionary Forces.

Following Dunkirk and a period of forced recuperation there began the costly and futile attempt to force England to its knees through massive application of air power alone. Since there were no really suitable aircraft for this undertaking – neither heavy long range bombers nor accompanying escort fighters – the German losses were correspondingly high. The attack strength dropped day by day. By the battle's end, British fighters and anti-aircraft units had shared in the destruction of a total of 184 Do 17, 20 He 59, 24 He 115, 225 He 111, 65 Ju 87, 271 Ju 88, 570 Bf 109, 243 Bf 110 and an additional 34 other types. Altogether, 1634 aircraft had fallen in the time period between 10 July 1940 and 31 September 1940. The Luftwaffe lost over 2000 men; 600 troops were wounded and another 630 taken prisoner in England. The Royal Air Force (RAF) or, to be more precise, Fighter Command, had lost a total of approximately 1300 aircraft, with the entire losses of the RAF amounting to over 1660 operational aircraft.

As a result of the dense radar coverage, the greater production of single-engine fighters (unlike Germany), and the unbearably high losses, the OKL was forced to reduce the number and strength of its attacks more and more.

Hitler had decided to attack the Soviet Union even before the air raids against British targets had begun. On June 22nd, 1941, German ground forces crossed over the border, supported from the air. Strategic bomber forces still had not been established, and even proposals to create a "Fernkampfkorps" (Long-range Bomber Corps) from the majority of bombers from the eastern Kampfgeschwader fell on deaf ears. As did suggestions for the formation of tactical Luftwaffenkommandos, whereby flak units and strike aircraft would effectively be able to work together.

The enemy's aircraft and armor factories were seldom touched, since they lay at the extreme edge of or outside of the Luftwaffe's range. In contrast, the German forces were plainly taxed to their limits by the vast expanse of territory, particularly since more capable aircraft were not available and the greater number of units were equipped with aircraft such as the Bf 109, the He 111 and the Ju 88. Once the advance had been stopped on the outskirts of Moscow, the enemy's armor and aviation forces grew increasingly larger. The severity of the costly winter battles, then the disaster of the 6th Armee in Stalingrad (1942), and finally Kursk (1943) led to ruinous losses for which the German aviation industry could not compensate. The shaken offensive units lacked the strength to turn the tide again. The Jagdwaffe was forced on the defensive more and more.

At the same time the Allied pressure over Western Europe and the Reich mounted. The production of heavy Allied bombers grew by leaps and bounds. And although day and night fighters were able to tally up impressive individual victories, the abandonment of long-range night fighter missions and the rare offensive thrusts by the Luftwaffe over England did little to change the situation. By 1943 the destructive power of the "four-engine heavies" was obvious to anyone. Cologne, Hamburg and Berlin are but three examples of the many cities which were struck. While at the same time jet aircraft development – initially put on hold after misinterpretation of the 1940 successes – required much more time than expected due to technical difficulties, particularly as a result of the problems caused by new engines.

Thus, on the 26th of June, 1944, the Luftwaffe certainly had a vastly superior number of operational aircraft than at the beginning of the war, but the fact that the units were equipped with obsolete operational aircraft types had changed relatively little. To be sure, a portion of the Jagdgeschwader had been equipped with the Fw 190 in the interim, and within the Kampfgeschwader the Ju 188 was replacing the Ju 88, yet at the same time barely more than 25 examples of the He 177 could be numbered among the forces in the West. In the East, the He 111 soldiered on alongside the Ju 87 and the Ju 88. The majority of the Jagdgeschwader continued flying the Bf 109; only a few Staffeln were able to utilize the Fw 190 A:

Luftflotte 3(Paris)

Stab/FAG 123	Toussus le Buc	1 Fw 190 and 2 Ju 88
4(F)/123	St. André	9 Bf 109
5(F)/123	Monchy-Breton	8 Bf 109
1(F)/121	Toussus le Buc	7 Me 410
II Fliegerkorps		
Stab/NAGr 13	Chartres	2 Bf 109 and 1 Fw 190
1/NAGr 13	Chartres	11 Bf 109 and Fw 190
3/NAGr 13	Laval	10 Bf 109 and Fw 190
III/SG 4	Clermont-Ferrand	52 Fw 190 total
IX Fliegerkorps		
3(F)/122	Soesterberg	7 Ju 188
6(F)/123	Cormeilles	6 Ju 88 and Ju 188
Stab/KG 2	Gilze Rijen	4 Ju 188
I/KG 2	Gilze Rijen	10 Ju 188
II/KG 2	Gilze Rijen	6 Ju 188
III/KG 2	Hesepe	6 Do 217
5/KG 76	Gilze Rijen	2 Ju 88
III/Kg 3	Hesepe	9 He 111
Stab/KG 6	Melun-Villaroche	2 Ju 188
I/KG 6	Melun-Villaroche	16 Ju 188
II/KG 6	Melun-Villaroche	2 Ju 188
III/KG 6	Melun-Villaroche	7 Ju 188
Stab/KG 30	Zwischenahn	3 Ju 188
I/KG 30	Leck	20 Ju 188
4 and 6/KG 51	Soesterberg	14 Me 410
5/KG 51	Gilze Rijen	7 Me 410
Stab/KG 54	Eindhoven	2 Ju 88
I/KG 54	Eindhoven	14 Ju 88
III/KG 54	Eindhovnen	13 Ju 88
I/KG 66	Montdidier	6 Ju 88 and Ju 188
E-Staffel IV/KG 101	St. Dizier	6 Ju 88
Stab/LG 1	Melsbroek	1 Ju 88
I/LG 1	Le Culot	13 Ju 88
II/LG 1	Melsbroek	13 Ju 88
I/SKG 10	Tours	19 Fw 190
X Fliegerkorps		
Stab/FAG 5	Mont de Marsan	
1(F)/5	Mont de Marsan	total of 15 Ju 290
2(F)/5	Mont de Marsan	
4(F)/5	Nantes	4 Ju 290
3(F)/123	Corme Ecluse	7 Ju 88
1(F)/ 129	Biscarosse	4 Bv 222

Stab/KG 40	Bordeaux-Merignac	9 Fw 200
1 and 2/KG 40	Toulouse-Blagnac	12 He 177
II/KG 40	Bordeaux-Merignac	12 He 177
7/KG 40	St Jean d'Angley	
8 and 9/KG 40	Cognac	total of 23 Fw 200

2 Fliegerdivision

1(F)/33	St. Martin	11 Ju 88 and Me 410
2/NAGr 13	Cuers	10 Bf 109 and Fw 190
2/SAGr 128	Berre	4 Ar 96
Stab/KG 26	Montpellier	2 Ju 88
II/KG 26(LT)	Valence	27 Ju 88
III/KG 26(LT)	Montpellier	22 Ju 88
Stab/KG 77	Salon	1 Ju 88
I/KG 77(LT)	Orange-Caritat	18 Ju 88
III/KG 77(LT)	Orange-Caritat	16 Ju 88
6/KG 77	Istres	8 Ju 88
4/KG 76 and 6/KG 76	Istres	7 Ju 88
Stab/KG 100	Toulouse-Francazals	1 Do 217 and 1 He 177
III/KG 100	Toulouse-Francazals	26 Do 217

4 Jagddivision

Stab/JG 1	St. Quentin-Clastres	3 Fw 190
I/JG 3	St. Quentin-Clastres	14 Bf 109
I/JG 5	Mons en Chaussée	14 Bf 109
II/JG 11	Mons en Chaussée	19 Bf 109
I/JG 301	Epinoy	13 Bf 109
Stab/JG 27	Champfleury	6 Bf 109
I/JG 27	Vertus	39 Bf 109
III/JG 27	Connantre	32 Bf 109
IV/Jg 27	Champfleury	31 Bf 109
Stab/NJG 4	Chenay	2 Bf 109
I/NJG 4	Florennes	38 Ju 88
III/NJG 4	Juvincourt	18 Ju 88
Stab/NJG 5	Hagenau	2 Bf 110
I/NJG 5	St. Dizier	15 Bf 110
III/NJG 5	Anties s. Laon	17 Bf 110

5 Jagddivision

Stab/JG 2	Creil	2 Fw 190
I/JG 2	Creil	16 Fw 190
II/JG 2	Creil	46 Bf 109
III/JG 2	Creil	18 Fw 190
Stab/JG 3	Evreux	3 Bf 109 and Fw 190
II/JG 3	Guyancourt	in transition
III/JG 3	Mareilly	23 Bf 109
II/JG 5	Evreaux	51 Bf 109
Stab/JG 11	Le Mans	3 Fw 190
I/JG 11	Le Mans	19 Fw 190
10/JG 11	Le Mans	9 Fw 190
I/JG 1	Alencon	17 Fw 190
II/JG 1	Alencon	5 Fw 190
Stab/JG 26	Guyancourt	3 Fw 190
I/JG 26	Guyancourt	27 Fw 190
II/JG 26	Guyancourt	12 Fw 190
Stab/NJG 2	Coulommiers	4 Ju 88
I/NJG 2	Chateaudun	10 Ju 88
II/NJG 2	Coulommiers	24 Ju 88
II/NJG 4	Coulommiers	13 Ju 88
II/JG 53	Vannes	32 Bf 109
1 to 3/JGr 200	Orange and Avignon	total 34 Fw 190
Stab/ZG 1	Bordeaux-Mérignac	1 Ju 88
1 and 3/ZG 1	Corme Ecluse	4 Ju 88
2/ZG 1	Chateauroux	10 Ju 88
III/ZG 1	Cazaux	7 Ju 88

In the spring of 1935 the first Dornier Do 23 left the factory in Friedrichshafen. the Do 23 G with BMW VIU engines served as the first bomber of the newly-formed Luftwaffe. In addition, this model was also used for many other purposes, including mine hunter, forestry and training aircraft.

Recruits being sworn in at the Lager-Lechfeld airfield. The airplane in the background is a Junkers Ju 86 D with 600 hp Jumo 205 C diesel inline engines.

Landing accident involving one of 6 Geschwader's He 111 E-3 bombers armed with three MG 15s. The aircraft was an asset of Luftkreiskomando V (Munich) and is seen in the camouflage scheme typical for 1938.

This photo was taken during a post-maintenance flight of an He 111 H. Callsign for this aircraft was DM+LE.

Do 217 E-4s of 6/KG 2 – here U5+AP – often took off from the airfield at Amsterdam-Shipol in May of 1942.

This Do 217 E-4 of 8/KG 2 also took part in night raids against England. Due to the concentrated use of British night fighters the operating crews suffered severe losses. The blackened callsign U5 +FS can be easily made out, as can the last two letters repeated on the rudder.

The successor to the Do 217 E series was the Do 217 K with a full-view canopy. Series production began at the end of 1942 for the Do 217 K-1 (here I/KG 66) at the Norddeutschen-Dornier-Werke in Wismar.

In addition to the Do 217 the Ju 88 was the standard bomber of the Luftwaffe. This Ju 88 A-1 of KG 30 has two additional MG 17 guns installed for side defenses. The cord running down the side of the fuselage is the release cable for the rubber liferaft.

For operational missions over water a number of Ju 88 A-4s of KG 6 were given a special camou-flage pattern, designed to break up the outline of the aircraft.

Due to the precarious supply situation with the BMW 801 engines, the Junkers-Werke proposed the Ju 188 A. The photo shows an A-2, which was used by I/KG 6 over England.

In the summer of 1944 the Geschwadern on the Eastern Front and in northern Europe also had a number of night strike planes in addition to the types previously mentioned. The aircraft used in this role primarily included the He 50, the Ar 66, the Go 145 and, in smaller numbers, the Ju W 34. For anti-tank operations and close air support the Schlachtgeschwadern (for the most part converted from the Stuka Geschwadern) often employed the Fw 190 F and G, the Ju 87 D-1 to D-7, and in lesser quantities, the Bf 109 equipped with an appropriate conversion kit. On 26 June 1944, the northern and eastern Luftwaffe units had the following aircraft:

Luftflotte 5 (Norway)

1(F)/124	Kirkenes	20 Ju 88 and Bf 109
1(F)/32	Kemijärvi	15 Fw 189 and Bf 109
3/(F)SAGr 130	Kirkenes	8 Bv 138
I/SG 5	Kirkenes	51 Fw 190 and Ju 87
III/JG 5	Petsamo	24 Bf 109
13(Zerstörer-staffel)/JG 5	Kirkenes	16 Bf 110

Luftflotte 1 (Latvia)

Stab/FAG 1	Riga-Spilve	
3(F)/22	Riga-Spilve	total of 7 Ju 188
5(F)/122	Mitau	9 Ju 88
Nachtstaffel 3	Riga-Spilve	15 Do 217
1/SAGr 127	Reval-Uleministe	12 Ar 96, Hs 126, Ju 60
14(Eis)/KG 55	Jakobstadt	9 He 111

3 Fliegerdivision

Stab/NAGr 5	Petersi	4 Bf 109 and Fw 189
1/NAGr 5	Idriza	13 Bf 109 and Fw 189
1/NAGr 31	Wesenberg	11 Fw 189 and Hs 126
II/SG 3	Jakobstadt	35 Ju 87 and Fw 190
Stab/NSGr 1 and 3 Staffel	Idriza	28 Go 145 and Ju W 34
1 and 2/NSGr 1	Kovno	in transition
Stab, 1 & 2/NSGr 3	Vecumi	49 Ar 66 and Go 145
1/NSGr 12	Vecumi	18 Ar 66
Stab, 1 & 2/NSGr 11	Rahkla	31 he 50 and Foller CV
2/NSGr 12	Libau	building up
Stab/JG 54	Dorpat	12 Fw 190
1/JG54	Turku(Finland)	Fw 190
2 & 3/JG 54	Reval-Laksberg	22 Fw 190

Gefechtsverband Kulme (Immola/Finland)

1/NAGr 5	Immola	Bf 109
II/JG 54	Immola	Fw 190
Stab/SG 3	Immola	Ju 87
1/SG 3	Immola	Fw 190
2 & 3/SG 3	Immola	Ju 87

Luftflotte 6

Stab/FAG 2 and		
4(F)/11	Baranovichi	2 Ju 188 and 3 He 111
4(F)/14	Baranovichi	15 Ju 188 and Do 217
NSt 4	Bobruysk	7 Ju 88 and Ju 188
14(Eis)/KG 3	Puchevichi	13 Ju 88
Stab/KG 1	Prowehren	1 Ju 88

IV Fliegerkorps

1(F)/100	Pinsk	14 Ju 88
Stab/NAGr 4	Piala-Podlaska	3 Bf 109 and Fw 189
3/NAGr 4	Kobryn	5 Hs 126
12/NAGr 3	Brest-Litovsk	8 Bf 109, Hs 126, Fw 189
Stab/KG 4	Bialystok	8 He 111 and Ju 88
II/KG 4	Baranovichi	34 He 111
III/KG 4	Baranovichi	40 He 111
Stab/KG 27	Krosno	1 He 111
III/KG 27	Krosno	41 He 111
III/KG 27	Mielec	35 He 111
Stab/KG 53	Radom	1 He 111
I/KG 53	Radom	36 He 111
II/KG 53	Piastow	37 He 111
III/KG 53	Radom	36 He 111
Stab/KG 55	Deblin-Irena	1 He 111
I/KG 55	Deblin-Ulez	35 He 111
II/KG 55	Deblin-Irena	35 He 111
III/KG 55	Grojek	36 He 111

1 Fliegerdivision

Stab/NAGr 4	Urezchye	2 Fw 189
I/NAGr 4	Bobruysk	12 Fw 189
11/NAGr 11	Urezchye	9 Fw 189
11/NAGr 12	Urezchye	7 Fw 189

Stab/SG 1	Pastovichi	5 Fw 190
III/SG 1	Pastovichi	38 Fw 190
I/SG 10	Bobruysk	10 Fw 190

4 Fliegerdivision

Stab/NAGr 10	Tolochin	3 Bf 109, Fw 189, Hs 126
2/NAGr 4	Orsha	18 Fw 189 and Hs 126
13/NAGr 14	Tolochin	7 Bf 109 and Fw 189
I/SG 1	Tolochin	44 Ju 87
II/SG 1	Vilna	73 Fw 190 and Ju 87
10(Pz)/SG 1	Bojari	20 Ju 87
10(Pz)/SG 3	Tolochin	22 Ju 87
Stab/SG 10	Dokudovo	3 Fw 190
III/SG 10	Dokudovo	39 Fw 190

Fliegerführer 1 (Minsk)

12/NAGr 12	Mogilev	12 Bf 109
2/NAGr 5	Budsslav	5 Bf 109
4/NAGr 31	Budsslav	6 Fw 190
Stab/NSGr 2	Lida	2 Ju 87 and Ar 66
1/NSGr 2	Bobruysk	14 Ju 87
3/NSGr 2	Lida	21 Ju 87
4/NSGr 2	Mogilev	17 Ju 87
1 Ostfliegerstaffel (Russia)	Lida	9 Go 145, Ar 66, U2
1 and 2 NSGr 1	Kovno	in transition
Stab/JG 51	Orsha	5 Bf 109 and Fw 190
Stab/StJG 51	Orsha	12 Bf and Fw 190
I/Jg 51	Orsha	35 Bf 109
III/JG 51	Bobruysk	31 Bf 109
IV/JG 51	Mogilev	17 Bf 109
III/JG 11	Dokudovo	19 Bf 109
Stab I/NJG 100	Baranovichi	total of 51 Ju 88
1 and 3/NJG 100	Biala-Podlaska	Do217 and Fw 190
4/NJG 100	Puchovichi	4 Ju 88

Luftflotte 4

2(F)/11	Jasionka	9 Ju 88
2(F)/22	Focsani	8 Ju 88
2(F)/100	Lublin	7 Ju 188

1 Fliegerkorps (Rumania)

Unit	Location	Aircraft
3(F)/121	Zilistea	8 Ju 88
Nst 1	Focsani	14 Do 217 and He 111
Stab/NAGr 1	Kishinev	3 Bf 109 and Fw 189
2/NAGr 16	Kishinev	10 Bf 109 and Fw 189
Stab and 1/NAGr 14	Comrat	15 Bf 109
2/NAGr 14	Bacau	15 Bf 109
Stab/FAGr 125(See)	Constanza	1 BV 138 and 2 Ar 196
1(F)/125(See)	Varna	9 BV 138 and Ar 196
3(F)/125(See)	Mamaia	8 BV 138 and Ar 196
Stab/JG 52	Manzar	1 Bf 109
I/JG 52	Leipzig(Rumania)	23 Bf 109
II/JG 52	Manzar	11 Bf 109
III/JG 52	Roman	19 Bf 109
15(Kroat.)/JG 52	Zilistea	in transition
Stab/SG 2	Husi	1 Ju 87
I/SG 2	Husi	29 Ju 87
II/SG 2	Zilistea	27 Fw 190
III/SG 2	Husi	43 Ju 87
10(Pz)/SG 2	Husi	16 Ju 87
II/SG 10	Culm	29 Fw 190
10(Pz)/SG 9	Trotus	15 Hs 129
14(Pz)/SG 9	Trotus	15 Hs 129
Stab/NSGr 5	Mazar	5 Go 145 and Ar 66
1/NSGr 5	Roman	21 Go 145 and Ar 66
2/NSGr 5	Kishinev	40 Go 145 and Ar 66
I/KG 4	Foscani	43 He 111

VIII Fliegerkorps (Poland)

Unit	Location	Aircraft
2(F)/11	Jasionka	9 Ju 88
2(F)/100	Lublin	7 Ju 188
Stab/NAGr 2	Strunybaby	5 Bf 109
1/NAGr 2	Stry	11 Bf 109
2/NAGr 2	Strunybaby	13 Bf 109
Stab IV(PZ)/SG 9	Lysiatycze	6 Hs 129
12(Pz)/SG 9	Stry	7 Hs 129
13(Pz)/SG 9	Lystiatycze	16 Hs 129
Stab/SG 77 and I/SG 77	Jasionka	in transition
I/SG 77	Jasionka	in transition
II/SG 77	Lemberg	33 Fw 190
III/SG 77	Cuniow	42 Ju 87
10(Pz)/SG 77	Starzawa	19 Ju 87
Stab/NSGr 4	Hordinia	4 Go 145

1/NSGr 4	Hordinia	28 Go 145
14(Eis)/KG 27	Krosno	15 He 111
7/NAGr 32	Labunia	11 Bf 109 and Fw 190

Commanding General of the German Luftwaffe in Rumania (Bucharest)

I/JG 53	Targsorul-Nou	28 Bf 109
III/JG 77	Mizil	28 Bf 109
10 and 12/NJG 6	Otopeni	25 Bf 110
11/NJG 6	Zilistea	7 Bf 110
2/NJG 100	Otopeni	8 Ju 88
4/JG 301	Mizil	6 Bf 109
6/JG 301	Targsorul-Nou	7 Bf 109

In addition to these, the Luftwaffe was also augmented by units from the Rumanian, Bulgarian, and Hungarian air forces, which were not only equipped in part with German piston-engine fighters, but also with relatively modern aircraft of their own design such as the Rumanian IAR 80 fighter.

The catapult-launched shipborne He 60 B-2 went into operation in 1930 as a reconnaissance aircraft and a B-class trainer for pilot, observer, radio and gunnery training.

Emergency landing of one of the few Ar 199 naval trainers, near Tournay in France.

An Ar 196 of Bordfliegergruppe 196 on the battleship "Scharnhorst" during a stopover in the military harbor at Brest. Aircraft callsign was T3+GH.

Beginning in 1937 the Heinkel company produced limited numbers of the catapult-launched He 114 A-2 maritime reconnaissance aircraft. The type was also later delivered to Rumania and Sweden.

Using a motor boat, a Do 18 N equipped for air-sea rescue operations is taken in tow to its berth. The two Jumo 205 engines were arranged in tandem.

The He 115 B-1 spent its service life primarily in the maritime reconnaissance, torpedo bomber and mine-laying roles. Its defensive armament consisted solely of two MG 15 machine guns.

Anchoring an He 115 to its buoy.

The military version of the Do 26 naval transport of Sonderstaffel Transozean with four tandem-mounted Jumo 205 E engines. The Do 26 V1 and V3 were lost in Norway on 28 May 940.

The Ju 52 was also given floats and converted to a naval transport. The floats had a length of 11.34 m and weighed 440 kg (with ice runners 552 kg). This version was given the designations – based on variant – of: Ju 52/3 mg5e, -7e, -10e and 14e.

The BV 238 V1 (RO+EZ) was, with a length of 43.5 m and a wingspan of 60.17 m, the largest airplane operated by the Luftwaffe. The only one completed was destroyed in an Allied strafing attack on the Schaalsee in 1944 just shortly after its maiden flight.

Already obsolete at the outbreak of war, the Heinkel He 59 was eventually put to use as a trainer and air-sea rescue plane. The photo shows an He 59 D-1 without armament and non-glazed nose.

Rescue training utilizing an He 59 C, a multi-seat naval aircraft converted in 1940 from the B-series.

Storing a three-man dinghy in its dorsal fuselage compartment on a Ju 88 A-1 of KG 51. Notice the release cable on the fuselage leading from the cockpit to the cover plate.

The standard two-man life raft of a Ju 87 with pressurized air bottle, bellows and drift anchor.

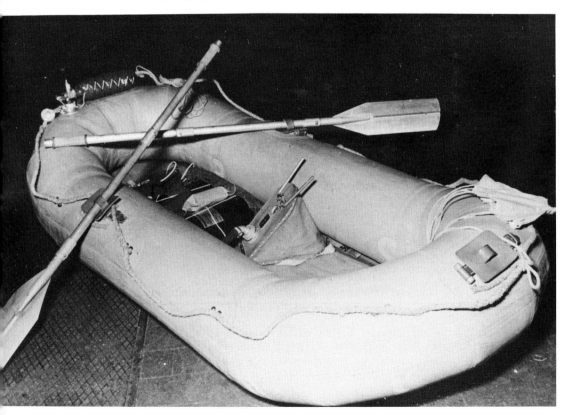

New Missions and Tactics 1944/1945

In August of 1944 the decision to reorganize the Luftwaffe's flying units was finalized. At the time, there were 9 1/3 Kampfgruppen equipped with the Ju 88 or Ju 188; these were to form the nucleus of 22 Kampfgruppen. By December of 1945 these were to have all been converted over to the Ju 388, plus the Me 262 and Ar 234.

All Do 217 bombers still being flown by KG 2 and KG 100 were to be phased out at an accelerated pace. Production was halted, with only a few further examples of this type coming from repair facilities. The same fate was in store for the Fw 200 (III/KG 49), the Do 217 "Kehl" aircraft (I, II/KG 2 and III/KG 100, and the He 177 A-3 and A-5 of II/KG 40 and II/KG 100). With the exception of those aircraft which were to be later handed over to an Erprobungskommando for remote-controlled weapons testing, these types, too, were to be phased out. The mass of the 37 He-177 "Kehl" bombers (with wire-guided Hs 293s) still operational and the 155 He 177 A-3/A-5 horizontal bombers were turned over as a reserve to the OKL in Denmark and Norway, where there was also a large stockpile of ordnance filled with poison gas.

KG 26, with its I and II Gruppe, operated as a torpedo unit in the summer of 1944 using the Ju 88 A-4/torp., Ju 88 A-17 and the Ju 188. It was planned that by the end of 1945 these Gruppen would be equipped solely with the Ju 388 M-1, thereby once again gaining the advantage in the engagement of maritime targets.

The He 111 units (III/KG 3 [with V 1], KG 4, KG 27 including 14/KG 27, KG 53, KG 55, and the Transportgruppe TG 30), too, were to only continue operations for a few months longer. Production on the He 111 was therefore stopped immediately and delivery left to those overhauled or repaired aircraft from the repair facilities. In August of 1944 there were roughly 180 He 111s left, plus 120 aircraft of the OKL reserve and 20 in use for pilot training (General der Ausbildung). Most of the remaining He 111 Geschwader units were to be converted over to the Me 262 as rapidly as possible. The OKL hoped to have the first two Gruppen operating the so-called Blitzbombers as early as September of 1944. By February of 1945 an additional 12 Gruppen would follow. But KG 4, 27, 53, and 55 never got the Me 262 A-2a, since its production had fallen far behind schedule. Along with this plan, there were also intentions to gradually equip these units with an improved version of the Ar 234, the C-3 and C-5, starting in July of 1945. A concept which never came to fruition.

Only KG 76 – and even then only II and III Gruppe – was reequipped with the Ar 234 B-2 starting in 1944. This was a result of the favorable operational performance which had been previously demonstrated by Kommando Lukesch. The planned production for March, 1945, of the Ar 234 C was terminated a short time later due to the progress of the war. The plan called for equipping 14 full Kampfgruppen with the Arado bomber.

Kampfgeschwader 2 was to be equipped with the Do 335 from the end of 1944. Some of the technical and maintenance personnel had already been transferred to Dornier to familiarize themselves with this modern airplane. Beginning in the spring of 1945 the RLM had anticipated a total of four fully operational Gruppen flying with the Do 335 A-1/A-2. Furthermore, in the late summer of 1944 it was still hoped that there would be two complete Kampfgruppen equipped with the Ju 287 A-1 by May of the coming year. In the interim, the Ju 88 K-1 would be phased into operations with Lehrgeschwader (LG) 1, KG 6 and the pathfinder unit, KG 66. The latter still flew the Ju 88 S-3 in the winter of 1944. The eventual plan called for 4 Do 335, 14 Ar 234, 2 Me 262 and 2 Ju 287 Gruppen. These were to be supplemented by two Gruppen of Ju 388 high-altitude bombers (LG 1), three Gruppen of standard bombers (KG 6) and a pathfinder Gruppe (KG 66).

Significant changes were also in store for the Zerstörer units. The two Zerstörergeschwader (ZG), 26 and 76, which were still in operation in the early fall of 1944 flew the Bf 110 and Me 410. Both units were to turn their current aircraft over by no later than February of 1945 and then begin conversion training on the Do 335 and the Ju 388, respectively. The OKL wanted to establish the first Do 335 Zerstörergruppe in April of 1945 and use it as a Mosquito hunter unit. In order to determine whether this type was suitable for such a role, there were plans to experimentally convert a number of Do 335s, a concept which went unfulfilled. In addition to eight Do 335 Zerstörer units, there were also to be an additional two Gruppen equipped with the high-performance Ju 388 J-1, replacing the Me 410 B-2/B-3 as quickly as possible.

By mid-1944 the Fernaufklärungsgruppen of the German Luftwaffe were primarily equipped with the Ju 88 D and the Ju 188 F. Altogether, there was a total of 21 Fernaufklärungsstaffeln with each having up to nine aircraft. Three Staffeln were equipped with the Me 410, which were to have been exchanged by October of 1944. As a result of serious production setbacks in Merseburg and other Junkers plants, however, several of these types were still in front-line operations as late as the spring of 1945. There were plans which called for equipping no less than three Staffeln with the Ar 234 B-1, the reconnaissance version of Arado's high-speed bomber. A Do 335 Staffel was also expected to undertake this role as well. At least one Do 335, the third prototype (CP+UC, WerkNr. 230003, later T9+ZH), was converted post-factory and operated by 1/Aufkl.Grp ObdL in the West during the fall of 1944. At the time there were up to five Staffeln operating night fighters – principally the Do 217. This number was eventually reduced to three Staffeln, which began gradually receiving the Ju 188 in September 1944.

Beginning in April, OKL's planning called for the conversion to a new standard model, the Ju 388 L-1. The Ju 290 would fulfill the role of long-range maritime reconnaissance, equipping 1 and 2/Fernaufklärungsgruppen (FAG). Future plans, however, anticipated a yet-to-be modified He 177 A-5 more suitable for the mission. The ultimate long-range goal was yet another conversion, this time to the Me 264. The two Bf 109 Nahaufklärungsstaffeln (tactical reconnaissance Staffeln), 4 and 5/(F) 123, would remain intact until the winter of 1945, at which time they would then also be disbanded.

By December 1945 there were to be three Staffeln with the Ar 234, ten units operating the Ju 388 as daytime reconnaissance and three as night reconnaissance, plus another thirteen with the Do 335.

In spite of all this careful planning the OKL was overtaken by the events of the war. The importance of the offensive bomber units, as with the long-range recce units, rapidly lost its significance. This was primarily due to a shortage of modern bombers in sufficient quantities as well as a lack of fuel. The fighter units had taken top priority in the interim. At the end of March 1945 the Geschwader were operating the following aircraft types:

Unit	Current Aircraft	Convert to	Comments
I/JG 1	He 162	no change	
II/JG 1	Fw 190 A-8 and A-9	He 162	
III/JG 1	Bf 109 G-10	He 162	(Apr-May 1945)
I/JG 2	Fw 190 D-9	no change	
II/JG 2	Fw 190 D-9	no change	
III/JG 2	Fw 190 D-9	no change	
I/JG 3	disbanded	—-	
II/JG 3	Bf 109 G-10	Bf 109 K-4	(phased in)
III/JG 3	Bf 109 K-4	no change	
IV/JG 3	Fw 190 A-8 and A-9	Fw 190 D-9	
I/JG 4	disbanded	—-	
II/JG 4	Fw 190 A-8 and A-9	Fw 190 D-9	
III/JG 4	Bf 109 K-4	no change	
IV/JG 4	Bf 109 G-10	Bf 109 K-4	(phased in)
II/JG 5	Bf 110 and Me 410	Bf 109 K-4	(Mar-Apr 1945)
III/JG 5	Bf 109 G-14	Bf 109 K-4	(phased in)
IV/JG 5	Bf 109 G-14	Bf 109 K-4	(phased in)
I/JG 6	Fw 109 A-8 and A-9	Fw 190 D-9	(phased in)
II/JG 6	Fw 190 A-8 and A-9	Fw 190 D-9	(phased in)
III/JG 6	Bf 109 G-14 As	Bf 109 K-4	(phased in)
I/JG 7	Me 262 A-1	no change	
II/JG 7	Me 262 A-1	no change	
III/JG 7	Me 262 A-1	no change	
Jagdgruppe 10	Fw 190 A-8, A-9 and D-9	assignment of type based on testing	
I/JG 11	Fw 190 A-8 and A-9	Fw 190 D-9	
II/JG 11	Bf 109 G-10	Bf 109 K-4	(phased in)
III/JG 11	Fw 190 A-8 and A-9	Fw 190 D-9	(phased in)
I/JG 26	Fw 190 D-9	Fw 190 D-12	(Apr 1945)
II/JG 26	Fw 190 D-9	Fw 190 D-12	(Mar-Apr 1945)
III/JG 26	disbanded	—-	
IV/JG 26	Fw 190 D-9	Fw 190 D-12	(Mar-Apr 1945)
I/JG 27	Bf 109 K-4	no change	1.98 boost increase
II/JG 27	Bf 109 G-10	Bf 109 K-4	(phased in)
III/JG 27	Bf 109 K-4	no change	1.98 boost increase
IV/JG 27	disbanded	—-	
Jagdverband 44	Me 262 A-1	no change	
I/JG 51	Bf 109 G-14	Bf 109 K-4	(phased in)
II/JG 51	disbanded	—-	
III/JG 51	Bf 109 G-14	Bf 109 K-4	(phased in)
IV/JG 51	Bf 109 G-14	Bf 109 K-4	(phased in)
I/JG 52	Bf 109 G-14	Bf 109 K-4	(phased in)
II/JG 52	Bf 109 G-14/U4	Bf 109 K-4	(phased in)

III/JG 52	Bf 109 G-14	Bf 109 K-4	(phased in)
I/JG 53	disbanded	----	
II/JG 53	Bf 109 K-4	no change	1.98 boost increase
III/JG 53	Bf 109 K-4	no change	1.98 boost increase
IV/JG 53	Bf 109 K-4	no change	1.98 boost increase
I/JG 54	Fw 190 A-8 and A-9	Fw 190 D-9	(phased in)
II/JG 54	Fw 190 A-8 and A-9	Fw 190 D-9	(phased in)
III/JG 54	Fw 190 A-8 and A-9	Fw 190 D-9	(phased in)
I/JG 77	Bf 109 G-14(U4)	Bf 109 K-4	(phased in)
II/JG 77	Bf 109 G-10	Bf 109 K-4	(phased in)
III/JG 77	Bf 109 G-10	Bf 109 K-4	(phased in)
I/JG 300	disbanded	----	
II/JG 300	Fw 190 A-8 and A-9	Fw 190 D-9	II-IV Gruppe
III/JG 300	Bf 109 G-10/R6	Bf 109 K-4/R6	conversion to
IV/JG 300	Bf 109 G-10/R6	Bf 109 K-4/R6	Me 262 A-1a at unspecified date
I/JG 301	Fw 190 A-9/R11	Fw 190 D-9/R11	later to Ta 152
II/JG 301	Fw 190 A-9/R11	Fw 190 D-9/R11	later to Ta 152
III/JG 301	Ta 152 H-10	no change	
IV/JG 301	disbanded	----	
I/JG 400	Me 163 B-1/B-2	He 162 A-1/A-2	conversion at
II/JG 400	Me 163 B-1/B-2	He 162 A-1/A-2	unspecified date
I/KG(J) 6	Bf 109 G-10/R6	Bf 109 K-4/R6	(phased in)
II/KG(J) 6	Bf 109 K-4	Bf 109 K-4/R6	(phased in)
III/KG(J) 6	Me 262 A-1	----	(reserved)
I/KG(J) 27	Bf 109 G-10/R6	Bf 109 K-4/R6	(phased in)
II/KG(J) 27	Bf 109 K-4	Bf 109 K-4/R6	(phased in)
III/KG(J) 27	Fw 190 A-9/R11	Fw 190 D-9/R11	(phased in)
I/KG(JJ) 54	Me 262 A-1	no change	
II/KG(J) 54	Me 262 A-1	no change	
III/KG(J) 54	Me 262 A-1	no change	
I/KG(J) 55	Bf 109 G-10/R6	----	I-III Gruppe to
II/KG(J) 55	Bf 109 K-4	----	be assigned to
III/KG(J) 55	Fw 190 A-9/R11	----	industrial site protection

The high losses suffered during the last weeks of the war, particularly among the new, hastily trained pilots and crews, did not fail to make an impact on the operational readiness of the units. What units remained were crippled on both the eastern and western fronts by constant retreats from one airfield to another, incessant attacks on dispersal areas and against aircraft landing or taking off, plus supplies and fuel going up in flames. Nevertheless, it was repeatedly possible to provide the ground forces with tangible relief. For example, Jagdgeschwader 6 (Horst Wessel) flew a total of 67 missions with 328 aircraft in the time space from 19 February to 18 March 1945. During this time, they succeeded in shooting down 20 single-engine and 4 twin-engine aircraft. Numerous ground strikes with the Bf 109 K-4 and Fw 190 A-9/D-9 destroyed 15 trucks. Despite these successes, the Geschwader lost six pilots, with two more pilots returning to base wounded. 25 single-engine fighters were either lost or damaged. A D-9 was shot down by friendly anti-aircraft fire near Cottbus. Ofw Naujoks lost his life. Together with JG 5, JG 52 and JG 77 also flew numerous low-level strikes in the Luftflotte 6 zone. Many of the remaining Staffeln fought their way back to lower Austria, the northern and southern Alpine lands and to the Prague area. All remaining units of the Luftwaffe were operationally organized as follows by the 3rd of May, 1945:

Unit	Remarks	Location
Luftflottenkommando 4	support of Heeres-gruppe Süd and Oberbefehlshaber Südost	Schöfling
18 Fliegerdivision	support of 6 SS Panzerarmee and 8 Armee	Wels
Nahaufklärungsgruppe 14	operational	Budweis
2/Nahaufklärungsgruppe 14	operational	Enns
3/Nahaufklärungsgruppe 14	operational	Budweis
2/Nahaufklärungsgruppe 16	operational	Budweis
1 Kette of 2 NAG 16	operational	Graz
Schlachtgeschwader 10	operational	Budweis
I/Schlachtgeschwader 10	operational	Budweis
II/Schlachtgeschwader 10	operational	Wels
IV (Pz)/Schlachtgeschwader 9	operational	Wels
13 (Pz)/Schlachtgeschwader 9	Panzerblitz tng.	Wels
14 (Pz)/Schlachtgeschwader 9	disbanding	Wels
II/Jagdgeschwader 52	operational w/o 1 Staffel	Hörsching
1 Staffel of II/JG 52	operational	Zeltweg
I/Jagdgeschwader 76	disbanding	Hörsching
II/Jagdgeschwader 51	disbanding	Hörsching
I/Jagdgeschwader 53	disbanding	Hörsching
Nachtschlachtgruppe 5	operational	Allensteig
1/Nachtschlachtgruppe 5	operational	Steinakirchen
I/Nachtschlachtgruppe 5	1 Staffel aux afld	Allensteig
2/Nachtschlachtgruppe 5	operational	Allensteig
3/Nachtschlachtgruppe 5	operational	Retz
2/Nachtschlachtgruppe 5	operational	Wels
17 Fliegerdivision	support for Oberbe-fehlshaber Südost, 2 Panzerarmee and 6 Armee	Bruck a.d. Mur
Nahaufklärungsgruppe 12	operational	Agram-Lucko
1/Nahaufklärungsgruppe 12	operational	Agram-Lucko
2/Nahaufklärungsgruppe 12	operational	Graz
Nahaufklärungsst. Kroatien	operational	Agram-Lucko

Unit	Remarks	Location
I/Schlachtgeschwader 2	operational	Graz
Nachtschlachtgruppe 7	disbanding	Agram-Gorica
1/Nachtschlachtgruppe 7	disbanding	Graz
2/Nachtschlachtgruppe 7	disbanding	Agram-Gorica
3/Nachtschlachtgruppe 7	disbanding	Agram-Gorica
8 Jagddivision		Wolfsleithein
4 Nachtjagdgeschwader 100	operational	Raffelding
Teilkommando 4 NJG 100	operational	Prossnitz
3(F)/Aufklärungsgruppe 121	under Lw Kdo 4	Hörsching
1(F)/Nachtaufklärungsgr.	under Lw Kdo 4	Hörsching
Fernaufklärungsgruppe 4	disbanding under Lw Kdo 4	Hörsching
2(F)/Aufklärungsgruppe 11	disbanding under Lw Kdo 4	Kirchham
3(F)/Aufklärungsgruppe 33	disbanding under Lw Kdo 4	Kirchham
III/Transportgeschwader 3	under Lw Kdo 4	Lindach
Luftkommando 8	support of Heeres-gruppe Mitte	Hermannstädtel
	support of Panzer AOK 4	Görlitz
3 Fliegerdivision	support of Panzer AOK 1	Olmütz
Nahaufklärungsgruppe 26	operational	Olmütz/West
1/Nahaufklärungsgruppe 26	operational	Olmütz/West
1/Nahaufklärungsstaffel 31	operational	Zauchtel
1/Nahaufklärungsgruppe 14	operational	Senoschat bei Deutsch Brod
Panzeraufklärungsschwarm 4	support of Panzer AOK 1	Kosteletz
Schlachtgeschwader 4	operational	Kosteletz
I/Schlachtgeschwader 4	operational	Kosteletz
III/Schlachtgeschwader 4	operational	Kosteletz
components of SG 4	formerly I and II/SG 4	Zauchtel
III/Schlachtgeschwader 10	operational	Prerau
Nachtschlachtgruppe 4	operational	Olmütz/South
2/Nachtschlachtgruppe 4	operational	Olmütz/South
Jagdgeschwader 77	operational	Prossnitz
I/Jagdgeschwader 77	operational	Peterswald
II/Jagdgeschwader 77	operational	Prossnitz
Jagdgeschwader 52	operational	Deutsch Brod
I/Jagdgeschwader 52	operational w/o 1/JG 52	Deutsch Brod
10(Pz)Schlachtgeschwader 9	operational	Deutsch Brod
Gefechtsverband Weiss	support of AOK 17	Schweidnitz

Unit	Remarks	Location
Nahaufklärungsgruppe 4	operational	Glaz-West
2/Nahaufklärungsgruppe 2	operational	Olmütz/West
1/Nahaufklärungsgruppe 12	operational	Schweidnitz
II/Schlachtgeschwader 77	operational	Schweidnitz
3/Nachtschlachtgruppe 4	operational	Ludwigsdorf
III/Jagdgeschwader 52	operational	Schweidnitz
1/Jagdgeschwader 52	operational	Alt Kemnitz
Gefechtsverband Rudel	operational	Niemes/South
Schlachtgeschwader 2	operational	Niemes/South
II/Schlachtgeschwader 2	operational	Niemes/South
10(Pz)/Schlachtgeschwader 2	operational	Niemes/South
I/Schlachtgeschwader 77	operational	Niemes/East
II/Jagdgeschwader 6	operational	Görlitz
Nahaufklärungsgruppe 15	operational	Reichenberg
1/Nahaufklärungsgruppe 15	operational	Reichenberg
2/Nahaufklärungsgruppe 15	operational	Reichenberg
12/Nachtaufklärungsst. 13	operational	Josefstadt/West
12/Nachtaufklärungsst. 13	auxiliary afld	Görlitz
Panzeraufklärungsschwarm 3	support of AOK 17	Reichenberg
Fernaufklärungsgruppe 3	operational	Königgrätz
2(F)/Aufklärungsgruppe 100	disbanding	Königgrätz
4(F)/Aufklärungsgruppe 121	operational	Königgrätz
1(F)/Aufklärungsgruppe 121	operational	Hohenmauth
4(F)/Aufklärungsgr. Nacht	operational	Hohenmauth
Absprung-Kdo 4(F) Nacht	operational	Prague-Ruszin
Schlachtgeschwader 77	operational	Pardubice
III/Schlachtgeschwader 77	operational	Pardubice
II/Schlachtgeschwader 4	operational	Königgrätz
III/Schlachtgeschwader 2	operational	Milovice nr. Prague
7(Erg)/Jagdgeschwader 1	operational	Liebenau
10(Pz)/Schlachtgeschwader 77	operational	Görlitz
III(Erg)/Jagdgeschwader 1	operational w/o 7(Erg)/JG 1	Görlitz
Jagdgeschwader 6	operational	Reichenberg
I/Jagdgeschwader 6	operational	Reichenberg
Verbindungsstaffel LwKdo 8	operational	Chrudim
Transportfliegerführer	operational	Prague-Ruszin
I/Transportgeschwader 3	operational	Chakovice
II/Transportgeschwader 2	operational	Sbraslavice
5/Transportgeschwader 2	operational	Primislau
6/Transportgeschwader 2	operational	Sbraslavice
7/Transportgeschwader 2	operational	Skutsch
III/Transportgeschwader 2	disbanding	Klattau

Unit	Remarks	Location
Gruppe Herzog	FKB-Staffel and components of Schleppgruppe 1	Hohenmauth
Kampfgeschwader 4	operational	Königgrätz
I/Kampfgeschwader 4	operational	Königgrätz
III/Kampfgeschwader 4	operational	Königgrätz
Luftwaffenkommando West	in transition and renaming to Lw-Division Nordalpen	Scheffau
7 Jagddivision	disbanding	Zell am See
Nahaufklärungsgruppe 13	minus aircraft	Bad Reichenhall
1/Nahaufklärungsgruppe 13	flying components at 1(F)/100	Hörsching
2/Nahaufklärungsgruppe 13	in consolidation	Bad Reichenhall
3/Nahaufklärungsgruppe 13	into one Staffel	Bad Reichenhall
1(F)/Aufklärungsgruppe	operational	Hörsching
Jagdgeschwader 27	Stab minus acft	Erdeinsatz
I/Jagdgeschwader 27	transition to Jagdgruppe 27 from	Zell am See
III/Jagdgeschwader 27	I and III/JG 27	and Salzburg
Jagdgeschwader 53	Stab minus acft Erdeinsatz	
components of JG 53	II, III, and IV/JG 53 (under JG 300), transition to Jagdgruppe 53	Prien
	ground components of II and IV/JG 53	Erdeinsatz
Jagdgeschwader 300	Stab minus acft	
II/Jagdgeschwader 300	II and III/JG 300	
III/Jagdgeschwader 300	in transition to Jagdgruppe 300	Ainring
(Erg)/Jagdgeschwader 1	one Staffel each JG 27's remaining flying components	Mühldorf and Erding
Nachtschlachtgruppe 1	flying components	Bad Aibling
Nachtschlachtgruppe 2	minus II/NSG 2	Pocking
Nachtjagdgeschwader 6	disbanding and transition to two Staffeln of 15 acft each	unknown
I/Nachtjagdgeschwader 11	disbanding and re-organizing as Einsatzstaffel NJG 11	unknown
Transportgruppe 30	subordinate to Luftwaffenkommando West	Salzburg
IX Fliegerkorps (J)	operational	Prague-Ruszin

Unit	Remarks	Location
JV 44	in transition and renaming to IV/ Jagdgeschwader 7	Salzburg-Mazglan
Kampfgeschwader(Jagd) 6	operational	Prague-Ruszin
III/Kampfgeschwader 6	operational	Prague-Ruszin
Kampfgeschwader(Jagd) 54	only ground comp.	Holzkirchen
I and II/KG(Jagd) 54	flying components	Prague-Ruszin
	ground components	Holzkirchen
Jagdgeschwader 7	ground components	Mühldorf
I and III/JG 7	flying components	Prague-Ruszin
III/JG 7	ground components	Mühldorf
Kampfgeschwader 51	flying components	Prague-Ruszin
I/Kampfgeschwader 51	flying components	Prague-Ruszin
I/Kampfgeschwader 51	ground personnel deployed	Prague-Ruszin

The following list shows just how little progress was made in the conversion of earlier bomber units to fighter units under the auspices of the Reichsverteidigung, or Defense of the Reich. In the meantime, training of the Kampfgeschwader(Jagd) units had come under the control of IX Fliegerkorps(Jagd), with initial plans calling for single-engine piston fighters to be superseded later by the Me 262 A-1 as the standard aircraft. But limited manufacturing capability hindered a large-scale conversion of these units to the twin-engine high performance fighter. The majority of available jet fighter units in the spring of 1945 showed quite a small number of stock on hand. Often these units – such as KG 6 – had only a handful of Me 262 A-1a fighters, and of these there were several on non-operational status due to missing engines or replacement parts. The units equipped with modern jet aircraft were, for the most part, just being established by the 28th of April, 1945:

Luftflotte REICH	Generaloberst Stumpf
1(F)/Aufklärungsgruppe 33 Einsatzkommando	Ar 234
1(F)/AufklGr 5	Ar 234
I/JG 1	He 162
10/NJG 11	Me 262
1(F)/AufklGr 123	Ar 234
Stab/NAGr 6	Me 262
1/NAGr 6	Me 262
2/NAGr 6	Me 262
Stab/KG 76	Ar 234
II/KG 76	Ar 234
III/KG 76	Ar 234
Stab/NAGr 1	Me 262
1/NAGr 1	Me 262
2/NAGr 2	Me 262

Luftflotte 6	GFM Ritter von Greim	
1(F)/100	Ar 234	
	Me 262	
Aufklärungskommando Me 262	Me 262	
JV 44	Me 262	renamed JG 7 "Salzburg'
Stabsstaffel JG 1	He 162	
Stab KG 51	Me 262	renamed KG 51 "Prag"
I/KG 51	Me 262	renamed KG 51 "Prag"
II/KG 51	——	renamed KG 51 "Prag"
13/EKG 1	Me 262	
Stab JG 7	Me 262	
I/JG 7	Me 262	
III/JG 7	Me 262	
I/KG(J) 54	Me 262	
Kommando Sommer	Ar 234	

During the last few days of the war many of these aircraft had to be blown up, due primarily to a shortage of fuel (J2) and the impossibility of deploying them elsewhere. The last operational Staffeln and the remnants of the earlier operational Geschwader units pulled back to the area of Prague at the beginning of 1945, from where they flew their final missions. Thus, a handful of Me 262s of III/KG(J) 6, together with seven aircraft from I/KG 51 flew as Gefechtsverband Hogeback (previously KG 76), conducting several fighter and ground strike sorties in Bohemia on 5 May 1945. As late as the 8th of May, Me 262s flew a daring low-level strike against Soviet tank wedge formations. Once all the remaining non-flying aircraft had been rendered unusable, the last four pilots lifted their jets into the air and eventually landed in American occupied territory. The Second World War had come to an end.

The Do 17 Z-7 "Kauz I" experimental night fighter was superseded by the Do 17 Z-10. The "Kauz II" was a three-seat night fighter armed with four MG 17s and an MG 151/20 in the nose, infra-red search device and a high intensity search-light.

A modified Do 217 E-2 (WerkNr. 0174) flew at the Dornier Werke in Friedrichshafen as the prototype for the Do 217 N-2 series.

The Ju 88 G-1 belonged to NJG 3, stationed in Grove in the spring of 1945.

A well-intentioned wish ("Come Back") for this crew of a Ju 88 G-6 night fighter with SN -2 radar, painted on the boarding hatch.

Although its use as a daytime escort fighter was limited due to its inadequate speed and maneuverability, the Bf 110 was given a new lease on life in the role of night fighter. This is a G-4 with SN-2 radar, with four MG 17s and two MG 151/20 cannons as standard weapons – armament which would later be enhanced considerably.

For increasing the effectiveness of friendly observation forces, Focke-Achgelis developed the single-seat Fa 330 "Bachstelze", a non-motorized tethered gyro kite. The design had a length of 4.47 m and a rotor diameter of 7.30 m, and could be disassembled in seven minutes.

Around September of 1939 work began on the two-seat Flettner Fl 282 reconnaissance platform and submarine hunter. The first ground tests took place in August 1941 with the V1 and later, on 30 October 1941, the Fl 282 V2 carried out the first free flight. Trials were conducted from August of 1942 without problems in Travemünde and several landings were made on the deck of the aircraft control ship "Greif."

Exercise demonstrating the recovery of a downed pilot using an Fl 282.

Two Flettner Fl 282s, the V21 (CI+TU) and the V23 (CI+TW), were equipped with extra seats be-
hind the engine. From June of 1944 the Fl 282 V21 was in Schweidnitz for partial overhaul.

The Fa 223 was the largest German helicopter built during the war. During a long distance flight from Delmenhorst to Berlin the Fa 223 V11 suffered an engine malfunction and was forced to land near a highway. The engine change was carried out on the spot.

Close-up photo of the canopy of an Fa 223 with lens mount, to which an MG 15 could be fitted.

Between 6 and 23 September the Fa 223 V1 took off from the Gebirgsjägerschule Mittenwald sports field and landed at 17 different sites up to an altitude of 2300 m (Dresdner Hütte), demonstrating to good effect its suitability for mountain operations.

During an exercise (29 September to 5 October 1944), the Fa 223 V16 carried an infantry field gun, supplies and ammunition up roughly 1800 m to a higher position on the Wörnergrat. In order to accomplish this the helicopter crew made 19 flights totalling 3 hrs 37 min. In comparison, transport using pack animals would have taken several days.

While rotary wing aircraft were able to carry out specialized transport missions, the transportation capacity of fixed wing aircraft (such as this Ju 52, probably the best known example) was markedly superior.

Off-loading a heavy SC 250 bomb from the spacious fuselage of a Ju 52.

The Ju 52 was also indispensable as a medevac airplane. In place of seats for personnel (R version), the aircraft was fitted with an S conversion kit including harnesses and stretchers for wounded soldiers.

One of the numerous Ju 52s used by the Transportgruppen of the Luftwaffe, using an E conversion kit to create a "crate carrier."

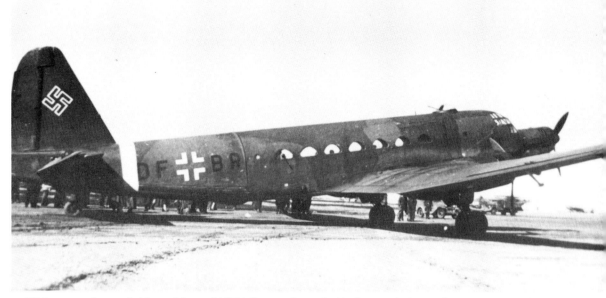

With a smooth metal skin and Jumo 211F inline engines, the Junkers Ju 252 was the successor to the legendary Ju 52. The V6 was one of 15 built in Dessau up until the end of 1941, before development was broken off due to the prevailing shortage of raw materials.

As an alternative solution, Junkers developed the Ju 352 in the fall of 1943 with wooden wings and fabric stretched over the steel tube fuselage. Three Bramo 323 R-2 engines provided 1000 hp each. By the time production stopped in August of 1944 Junkers had completed 44 additional aircraft in its branch plant at Fritzlar.

In the summer of 1937 Junkers produced the Ju 290 for Lufthansa. At the outbreak of the war these aircraft were appropriated for military applications, as illustrated by Ju 90 V7 (ex "Baden" D-ADFJ, WerkNr. 900003, later callsign GF+GH).

The Ju 90 V6, WerkNr. 4918, was put to use on the Eastern Front following its maiden flight on 30 July 1940.

Even the Focke-Wulf Fw 200 C passenger liner was to be found in Luftwaffe colors. This machine serves the Reich government as a comfortable travel plane.

Refuelling an He 100 D-1. After the fighter design was rejected, the aircraft was posed with a few others for propaganda photos and painted with imaginary unit markings.

Refuelling a Bf 109 E-3 in the field.

Using a hand pump, a ground crewman feeds the fuel from a fuel container into the fuselage tank of a Bf 109 D-1 of 2/JG 71. Fuel bowsers were not always to be found in sufficient quantity.

A Ford V3000 with fuel trailer, as seen in many of the Bomberstaffeln in the West.

An Arado Ar 68 E on its back. The aircraft probably belonged to JG 26 "Schlageter." It was powered by the high-performance Jumo 210 Da inline engine.

Unusual parking spot for a Bf 109 C-1

This Ju 52 stood in the way of a landing He 177 and had to be written off as a total loss.

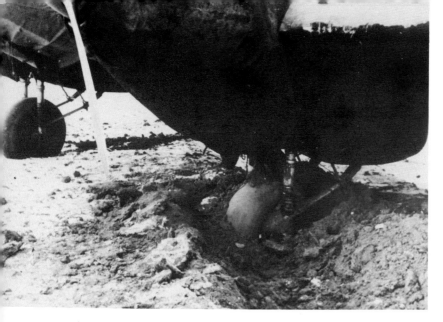

Landing accident of an He 111 H-4 which overshot the runway. The operational aircraft eventually came to rest in the soft earth, only slightly damaged.

Following a night mission over central England, this Ju 88 A-5 of KG 30 flipped over due to poor runway conditions.

This Ju 88 A-4 of LG 1 in Italy went out of control, but was finally brought to a halt by these two corrugated tin barracks.

During a low-level flight over the ocean the propeller blades of this Ju 88 D-1 long-range reconnaissance aircraft caught the water's surface and broke off.

For repairing the landing gear and changing out the BMW 801 engine, mechanics at a Feldwerft-Abteilung raised this Do 217 K-1 of I/KG 6 using inflatable air bags.

Close-up detail shot of the left engine nacelle with DB 603 A engine of a Do 217 N-2 following a belly landing.

A parked Ju 87 R-1 with 300 l drop tanks under its wings. Without these tanks, the aircraft was basically the same as the B-1 series.

A poorly camouflaged Do 17 M of KG 2, particularly since there are no other trees to be seen on the airfield.

A Bf 110 G-4 hidden down to its antenna equipment.

A well concealed He 177 A-3 of 8/KG 100 in Fassberg. Only the propellors have yet to be covered with camouflage netting.

In 1940 comprehensive testing began in the öztaler Alps with an Fi 156 using skis for landing gear.

An He 111 H-1 with skis in Norway, 1941. The aircraft belonged to KG 100 and carried the tactical callsign 6N+AX.

This Ju 87 D (D-INRE) was fitted with skis in place of its wheeled landing gear. The special gear was bolted directly to the wheel forks.

A Bf 109 F, WerkNr. 8195, neatly covered with a canvas tarp. Unusually, on this aircraft the undercarriage legs are shrouded.

In the air the fastest fighter of its day, but on the ground it needed a team of oxen to get around.

An alternative method is also being used to tow this Go 242 A-2. The aircraft carried a braking parachute of 5.0 m diameter for landing on smaller fields.

Loading a Ju 52 of KGrzbV 400 in the harsh Russian winter using panje sleighs.

An Fi 156 C-3 liaison aircraft of 4(F) 121. Provisions for the airfield personnel are permitted to roam the airbase freely.

A production He 111 getting its compass calibrated. The sheep in the foreground help to keep the grass down.

A Bf 110 of Aufklärungsgruppe (F) 14 during operations in North Africa.

In northern Germany at the war's end, cows graze among Ju 88 bombers which have had their rudders removed.

Photo Credits:

Also from the publisher

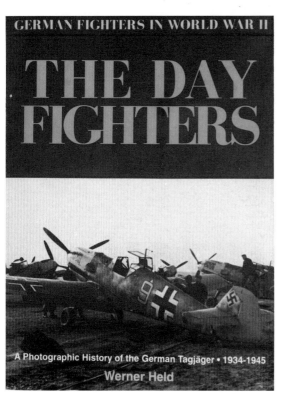

GERMAN FIGHTERS IN WORLD WAR II

THE DAY FIGHTERS

A Photographic History of the German Tagjäger • 1934-1945

Werner Held

Messerschmitt
Bf 110
OVER ALL FRONTS
1939-1945

Holger Nauroth / Werner Held

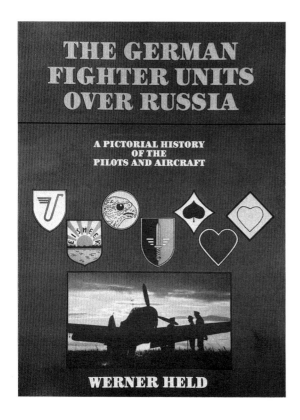

THE GERMAN FIGHTER UNITS OVER RUSSIA

A PICTORIAL HISTORY OF THE PILOTS AND AIRCRAFT

WERNER HELD

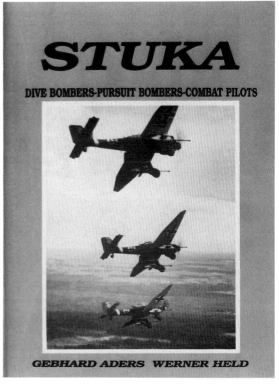

STUKA
DIVE BOMBERS-PURSUIT BOMBERS-COMBAT PILOTS

GEBHARD ADERS   WERNER HELD

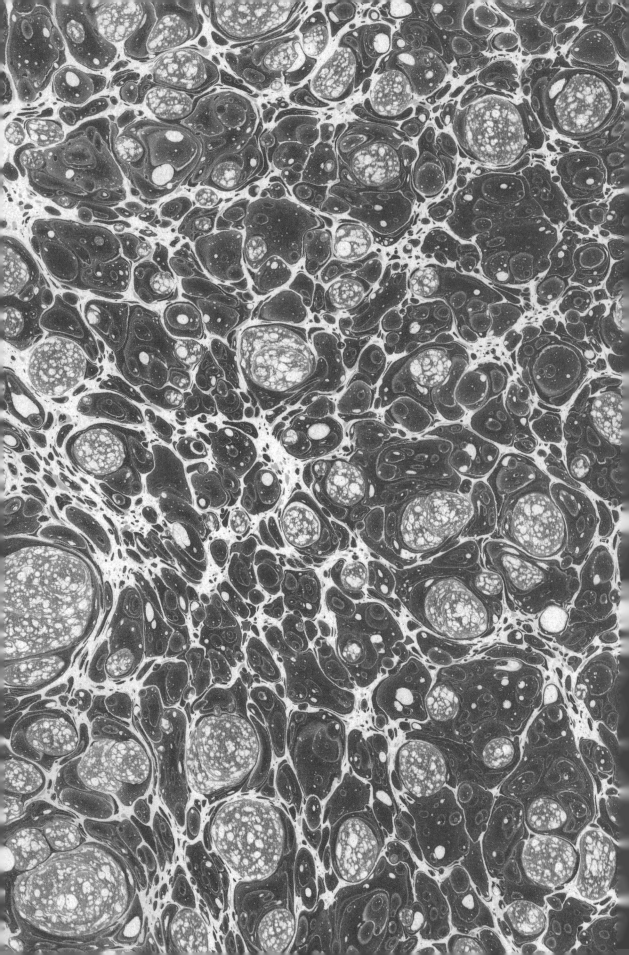